PRIMATES
IN·QUESTION

PRIMATES
IN·QUESTION

THE SMITHSONIAN ANSWER BOOK

ROBERT W. SHUMAKER · BENJAMIN B. BECK

PHOTOGRAPHS BY GERRY ELLIS

SMITHSONIAN BOOKS
WASHINGTON AND LONDON

© 2003 Smithsonian Institution

Photographs © 2003 by Gerry Ellis, except pages 53, 76, 95, 108, 123, 131, and 139

Copy Editor: Robin Whitaker
Production Editors: Ruth G. Thomson and Joanne Reams
Designer: Janice Wheeler

Library of Congress Cataloging-in-Publication Data
Shumaker, Robert W.
    Primates in question : the Smithsonian answer book / Robert W. Shumaker; Benjamin B. Beck ;
photographs by Gerry Ellis.
        p. cm.
  Includes bibliographical references (p.).
  ISBN 1-58834-151-8 (alk. paper) — ISBN 1-58834-176-3 (pbk. : alk. paper)
       1. Primates—Miscellanea.   I. Beck, Benjamin B.   II. Title.
  QL737.P9S466   2003
  599.8—dc21   2003042463

British Library Cataloguing-in-Publication Data available

Printed in China, not at government expense
10 09 08 07 06 05 04 03   1 2 3 4 5

♾ The paper used in this publication meets the minimum requirements of the American National
Standard for Information Sciences—Permanence of Paper for Printed Library Materials ANSI
Z39.48-1984.

Frontispiece: Gibbon (*Hylobates* sp.)

CONTENTS

Preface   ix
Acknowledgments   xi
Introduction   xiii

.1.
PRIMATES IN GENERAL

What Are Primates?   1
How Are Primates Classified?   1
What Are the Different Types of Primate?   4
What Makes a Primate a Primate?   36
Are Monkeys and Apes the Same Thing?   42
How Do Primates Move?   45
How Closely Are Humans Related to Other Primates?   52
Did Humans Evolve from Apes?   54
Will Chimpanzees Evolve into Humans?   56
What Are the Smallest and Largest Primates?   57
What Do Primates Eat?   59
Do All Primates Have Tails?   61
How Long Do Primates Live?   65
Which Primates Are the Fastest?   66
How Strong Is a Gorilla?   67

.2.
PRIMATE SOCIAL BEHAVIOR

Do All Primates Live in Families?   71
Why Do Primates Live in Different Types of Groups?   72
Are Primates Monogamous?   75
Do All Primates Recognize Their Relatives, and Does This Influence
  Their Society?   77
How Does Body Size Affect the Lives of Primates?   79
How Long Do Babies Stay with Their Mothers?   81
Do Mothers Have Help in Raising Their Babies?   83
Do Monkeys and Apes Experience Adolescence?   86

Why Do Some Primates Have Swollen Rears?   88

When Do Primates Mate?   90

Do Males or Females Initiate Mating?   91

Will Primates Adopt a Baby?   93

How Have Ideas about Primates Changed?   96

Why Do Primates Spend So Much Time Grooming?   97

Which Primates Hunt for Meat?   100

Do All Primates Have Friends?   102

Do Primates "Make Up" after a Fight?   103

Do Primates Make Good Pets?   105

.3.
PRIMATE INTELLIGENCE

How Smart Are Primates?   109

How Do Scientists Study Primate Intelligence?   109

When Was Primate Intelligence First Studied?   111

What Is the Average IQ for an Ape?   114

Do Primates Have Big Brains?   116

Do Nonhuman Primates Use Tools?   118

Do Nonhuman Primates Make Tools?   120

How Do Primates Communicate?   124

Do All Primates Have Their Own Languages?   127

Can Nonhuman Primates Learn a Language?   128

Why Can't Apes Talk?   131

Can Monkeys and Apes Count?   132

Do Nonhuman Primates Make Up New Ways to Solve Problems?   134

How Do Primates Learn from Each Other?   136

Which Primates Have Culture?   139

Are Humans the Only Deceptive Primate?   142

Are All Primates Sympathetic?   145

Which Primates Have Emotions?   146

.4.
PRIMATE CONSERVATION

How Many Species of Primate Are Threatened or Endangered in the Wild?   151

What Are the Main Threats to Primates in the Wild?   153

What Are the Best Ways to Save Primates in the Wild?   157

Who Is Working to Protect Primates?   158

Are Captive Primates in Trouble?   161
How Can I Get Involved?   163

Appendix: Taxonomic Hierarchy of Primates   165
Glossary   175
References   179
Taxonomic Index   191
Subject Index   193

Photo gallery appears following page 64.

PREFACE

Primates represent many different things to many different people. They may be revered as sacred symbols or considered pests that steal crops. Some treat them as surrogate children, and others use them to further biomedical research. Increasingly, many fill their cooking pots with primates, while conservationists attempt to preserve remaining populations and their habitat. Many live in captivity; fewer and fewer survive in the wild.

The blessing and curse for nonhuman primates are the set of close similarities they share with us. These characteristics can both attract and repel, leading to the remarkable range of ways in which humans perceive our closest evolutionary relatives. This book focuses primarily on the work of scientists, researchers, dedicated conservationists, and others that share a mutual attraction and devotion to the 350 different species that constitute the order Primate.

The first chapter deals with the anatomy, taxonomy, and general biology related to understanding primates. We have attempted to utilize the most current and well-informed opinions regarding the classification of the order; however, it is inevitable that new information will continue to revise our understanding of primate taxonomy. Although occasionally frustrating, the process of science is always self-correcting. The second chapter of the book describes many of the common social behaviors that occur among primates. Most species are intensely social by nature, and humans are no exception. It is very likely that readers will see themselves reflected more than once in the information found in this portion of the book. The third chapter is devoted to the mental abilities that have been documented for primates. Information presented here may be surprising to some, and a few common misperceptions are dispelled. The conclusions drawn in this chapter deal with a relatively small number of primate species, and further research will greatly increase our understanding of the cognitive ability that exists throughout the order. Conservation and welfare are addressed in the final chapter. It is no exaggeration to say that the situation for primates in the wild has never been as precarious as it is today. A listing of organizations that actively promote the interests of primates in the wild and in captivity is provided so that the reader can easily acquire more information. Finally, an appendix includes a complete listing of all primate species, and a reference section supports the information that is presented throughout the book.

Squirrel monkey (*Saimiri* sp.)

ACKNOWLEDGMENTS

In any field of science, major discoveries by lone individuals are the exception. As a rule, progress in science is incremental and collaborative by nature. This book shares those two distinctions, having progressed incrementally, with the final product benefiting immensely from the assistance of many colleagues. While we hope we have included everyone who provided expertise and advice, apologies are extended to anyone who may have unintentionally been forgotten.

John Fleagle, Susan G. Larson, and Ted Grand offered clear answers to complex questions about anatomy and movement. Andy Baker instantly summoned citations to deal with very specific aspects of golden lion tamarin behavior. Karyl Swartz clarified murky areas of the psychology literature. Ann M. Palkovich lent her expertise in physical anthropology, and Brad Blaine proved his mastery of both Greek and Latin. Charlene Jendry and Tony Murakikwa shared important insights about gorilla behavior. Christina Ellis was a tremendous source of information about the bushmeat trade in Africa.

Special appreciation is in order for Lynn Dolnick, Kathleen Samiy, and especially Elena Lopez. Thanks are due Caisie Pitman for her invaluable assistance. Tetsuro Matsuzawa and Richard Nowitz in particular were amazingly generous with their photographic contributions.

Colin Groves must be singled out for special praise. Although he was literally on the other side of the world with his own responsibilities and academic pursuits, it appeared that nothing was more important to him than answering my all-too-frequent e-mail messages about primate taxonomy. His expertise is staggering, exceeded only by his generosity and affability. Thank you very much indeed.

The last set of individuals to be recognized are the many nonhuman primates that I have been fortunate to know. Sylvia, Azy, Tomoka, Indah, Gustav, Tucker, Lucky, Bonnie, and the list includes many others. Each has been a source of inspiration, making me feel humbled and exhilarated simultaneously.

The royalties that I accrue from this book will be devoted entirely to primate research, education, and conservation.

This book is dedicated to Anne and William, my motivation.

R. W. S.

INTRODUCTION

For the most part, those of us who have grown up and live in temperate climates have never seen primates in the wild. Even in countries with populations of wild primates, few local people venture into the forest except to hunt. Some species of monkey coexist with humans in developed areas, but interactions in these circumstances are usually competitive and antagonistic. Although the average person is generally fascinated by primates, few people have the opportunity to learn much about them. At best, impressions are generally shaped by visits to good zoos or by high-quality wildlife documentaries. At worst, perceptions are formed by watching primates in "entertainment" that are forced to behave like degenerate people.

Those of us who work with primates for a living too often lose our perspective. We may cringe when someone calls a chimpanzee a monkey or asks, "Which type of primate makes the best pet?" We simply forget that most people have very little access to good information about primates. In addition to caring for, studying, and protecting primates in the wild and in captivity, we also need to be teachers. This book is an attempt to provide accurate and compelling information about primates to interested readers (and maybe convert some uninterested ones as well).

Primates are simply captivating in their appearance, behavior, mental abilities, and overall complexity. No other type of animal shares the same level of similarity with humans or has the same depth to their gaze. We know what it is to be a primate by our very existence. While differences between humans and other species of primate surely exist, there are remarkable, undeniable similarities as well. Human and nonhuman primates see the world through the same eyes and feel it through the same hands and feet. Primates know the value of social relationships, alliances, and cooperation. We have all been stung by competition and have suffered social rejection. Our shared biological heritage compels us, more than any other species, to invest in our mates, our offspring, and the success of our family groups.

At this point in our history, the other 349 species of primate are in dire need of our assistance, compassion, and conservation. A better understanding of the other primates will, it is hoped, lead to an increased desire to participate in their preservation and protection.

.1.

PRIMATES
IN GENERAL

WHAT ARE PRIMATES?

Primates are members of the taxonomic order Primate, a subgroup of mammals (class Mammalia). All primates have hair, regulate their body temperatures internally (are warm-blooded), give birth to live young, and feed their offspring with milk that is produced by the mother. In general, all primates have a shared set of physical features. The eyes are forward facing, resulting in accurate depth perception, and color vision is typical. Primates have two arms and two legs, rather than four legs as in most other mammals. They have hands and feet (not paws), most species have nails rather than claws, and an opposable set of digits is always present. Fingerprints are easily seen and can be used for individual recognition. Most primates give birth to singletons, which are nursed on two pectorally located breasts. Offspring have a long period of dependence and relatively slow rates of overall maturation. All primates have large brains relative to their body sizes. Perhaps more than any other type of animal, primates have captured the interest, imagination, and attention of humans for centuries. Given that we share all of these features and are also members of the order Primate, our fascination is understandable.

HOW ARE PRIMATES CLASSIFIED?

Primates, like all living things, are classified according to how closely or distantly they are related to each other. This system of classification is called taxonomy and was introduced by Carolus Linnaeus in the eighteenth century. Taxonomic classifi-

Orangutan (*Pongo* sp.)

This adult male chimpanzee exhibits many of the physical features that characterize all primates.

cation is an indispensable means for ordering the natural world by understanding the biological relationship that exists among different groups of organisms.

The degree of relatedness among groups is determined by studying physical features. Groups with shared characteristics that have been inherited from a common ancestor are more closely related than groups with similar characteristics that have evolved independently. Teeth are an especially useful example for illustrating this point. All living primates share certain dental features, such as the general shape and arrangement of the teeth in the jaw. These features have been inherited from a common primate ancestor, and a physical anthropologist can instantly identify a tooth that has come from a primate. Other groups of mammals, such as carnivores, obviously also have teeth, but their general shape is dramatically different and inherited from a distinct lineage. The taxonomic classification of primates and carnivores reflects the fact that these two groups share only the most general physical features and are clearly not closely related.

The taxonomic system for keeping track of how each organism is related to all others is termed binomial nomenclature, which is similar to our system of assigning a specific call number to all books in a library. That is, the specific scientific names given to every type of organism are internationally recognized and used by all scientists. Scientific names are composed of two words that generally provide some description of the organism. An easy example is the domestic dog, *Canis familiaris*. The first word names the genus (which is always capitalized), and the second names the species (which is never capitalized), and both are usually written in italics or underlined. Common names, such as "domestic dog," might vary in different languages, but scientific names are always the same. Therefore, scientists from any country can accurately refer to any specific organism without confusion.

Scientific names are derived from a system called the Linnaean hierarchy. This system includes seven different levels of categorization that are arranged from most general to most specific, as the following shows:

Kingdom—delineates the broadest categories, such as animal, plant, fungus, and bacterium

Phylum—begins differentiating groups within each kingdom, such as vertebrates and invertebrates among animals

Class—distinguishes organisms within a particular phylum, such as mammals and birds among vertebrates

Order—identifies organisms within a particular class, such as primates, carnivores, elephants, and rodents, among mammals

Family—identifies groups that are closely related within each order, such as the hominids among primates: gorillas, chimpanzees, bonobos, orangutans, and humans

Genus—arranges family members into more closely related groupings, such as orangutans among the hominids

Species—arranges members of the same genus, such as the orangutans, into the most closely related groupings: Bornean orangutans and Sumatran orangutans

In general, the species level of classification is used to identify a group of organisms that commonly, or potentially, reproduce with each other under natural circumstances. It is important to understand that this designation is based on the situation that normally exists for a group of organisms in the wild, where interbreeding can be prevented simply by behavioral differences or geographical boundaries. Closely related species that would never encounter each other in the wild have been able to reproduce in captive settings, but this does not change their status as different species.

The Linnaean hierarchy for Bornean orangutans and humans demonstrates our close evolutionary relationship:

Kingdom	Animalia	Animalia
Phylum	Chordata	Chordata
Class	Mammalia	Mammalia
Order	Primate	Primate
Family	Hominidae	Hominidae
Genus	*Pongo*	*Homo*
Species	*pygmaeus*	*sapiens*

Orangutans, the Asian great apes, are found only on the islands of Borneo and Sumatra. A young Bornean individual (*Pongo pygmaeus*) is pictured here. The geographically separate Sumatran orangutans, known as *Pongo abelii*, are a different species.

The scientific name for each group of organisms can be derived from the chart. Humans are known as *Homo sapiens*, and Bornean orangutans are known as *Pongo pygmaeus*.

As in all other areas of science, different taxonomists can hold varying views on certain issues. For example, taxonomists may disagree on the exact number of genera and species within the primate order. Remember that the genus level of classification is broader than the species level, and therefore one genus can include multiple species. This book follows the taxonomic system described in *Primate Taxonomy*, authored by Colin Groves, which classifies primates as having 350 distinct species.

WHAT ARE THE DIFFERENT TYPES OF PRIMATE?

All primates share a general set of anatomical features that can be used to distinguish them from the other mammals. The current level of classification, as listed in

Groves's *Primate Taxonomy*, recognizes 350 species of primate, which are related to each other in varying degrees. The evolutionary "distance" between these species is determined by studying a variety of physical characteristics. In any area of scientific research, active inquiry and the development of new analytical techniques may reveal information that changes traditional views, and this has certainly been true in the process of understanding and categorizing the different types of primate. Taxonomic classification is an ongoing and dynamic field of science that blends academic expertise in the disciplines of anatomy, genetics, evolutionary biology, and microbiology, among others. Over time, our understanding of primate taxonomy will doubtlessly be refined. The most widely accepted current opinion divides the primate order into two suborders, called the Strepsirrhini and the Haplorrhini. The strepsirrhine group includes all of the prosimians. Tarsiers, monkeys, and apes constitute the haplorrhines. The following sections, which are summarized in the appendix ("Taxonomic Hierarchy of Primates"), characterize the different taxonomic groups that make up the order Primate.

Suborder Strepsirrhini

The strepsirrhines are the "wet-nosed" primates. The species in this suborder have a moist nose that is fused with the upper lip, which is attached to the gum, resulting in a face that has a relatively limited range of expressions. All of the strepsirrhines are prosimians, a general-use term that translates as "pre-monkey," which describes the Lemuriformes, Chiromyiformes, and the Loriformes.

The prosimians are considered less complex than monkeys and apes, having characteristics that are primitive by comparison. They have reduced braincases with a corresponding decrease in absolute brain size. The olfactory lobes of the

This bushbaby is one of the "wet-nosed" prosimians, an evolutionarily more primitive type of primate. The shiny, moist skin of the nose can easily be seen.

Ring-tailed lemurs, like all prosimians, have an elongated snout and regularly use scent to communicate with each other. Their feet have a strong grip, and the big toe is extremely useful for holding, grasping, and climbing.

brain are large, indicating a greater reliance on the use of smell compared with the haplorrhines. An elongated snout is present, which gives the face a doglike appearance, although this is also the case for some monkeys, such as the baboons. All of the prosimians (except for the aye-aye, *Daubentonia madagascariensis*) have tightly clustered incisors and canine teeth in the mandible that form a "dental comb," which is used for grooming. The feet have two distinct anatomical specializations. The big toe is widely separated from the other toes, allowing for a very secure, viselike grip during locomotion, and the second toe usually has a distinctive grooming claw.

Strepsirrhine skulls have large orbits for the eyes, which are protected by a circular, bony bar. All have relatively big eyes, and on the back of the retina the majority have a highly reflective layer called the tapetum lucidum cellulosum. *Tapetum lucidum* is translated in Latin as "shining layer." This surface presumably acts to intensify light that comes into the eye and is more common in the nocturnal prosimians, which make up approximately 75 percent of the suborder. However, this structure is found in some diurnal (day-active) species as well, such as the ring-tailed lemur (*Lemur catta*). The tapetum lucidum cellulosum is responsible for the glowing appearance of the eyes of many animals when light is shined directly into them. Many of the nocturnal species also have very sensitive hearing and the ability to move each ear independently in order to capture sound more precisely.

Strepsirrhine reproduction has distinctive elements as well. All prosimians have a breeding season rather than individual cycles, and most have litters of offspring as well as multiple nipples for nursing. All have a bicornuate uterus, meaning it has two chambers that form a Y shape. During pregnancy, the blood vessels of the mother and offspring remain completely separate in a condition known as epitheliochorial placentation.

Many of the individual features that are present in the strepsirrhines also exist for other animals. It is important to remember that the unique blend of these physical characteristics is what distinguishes the prosimians within the primate order and points to their more primitive level of development compared with the haplorrhines. The following sections characterize the groups of species that constitute the strepsirrhines. This suborder includes the lemurs, lorises, pottos, and galagos.

Infraorder Lemuriformes
Families Cheirogaleidae, Lemuridae, Megaladapidae, and Indridae
Madagascar, the large island off the east coast of Africa, is the only place where lemurs are found in the wild. These prosimians range in size from the tiny *Microcebus*, weighing only 30.6 grams, to the *Indri*, which may reach up to 10 kilograms. Behavioral variation among the lemurs is clear and is especially obvious in the wide range of foods that they consume. While some species feed exclusively on vegetation, others consume everything from fruit to insect larvae. Most lemurs are vertical clingers and leapers (see *How Do Primates Move?*).

Infraorder Chiromyiformes
Family Daubentoniidae
The aye-aye (*Daubentonia madagascariensis*) is the most distinctive of the lemurs and the only species in the infraorder Chiromyiformes. It lives a solitary, nocturnal existence and survives largely on insect larvae as well as some fruits, flowers, vegetation, nuts, and seeds. Unlike any other primate, the aye-aye has rodentlike incisors, which it uses to gnaw away wood in search of insect larvae. It then extracts the larvae using the claw on an exceedingly long, bony third finger, which is present on each hand. It locates its prey by tapping on trees with the tip of this skeletal digit and listening for the movement of the prey in response. It has been suggested that the aye-aye occupies the ecological niche that woodpeckers occupy in their ranges, which don't include Madagascar. This specialized finger is also utilized to draw the contents out of raw eggs and coconuts. The aye-aye inserts the finger and pumps the contents out onto its tongue.

The ring-tailed lemur (above left), black-and-white ruffed lemur (above right), and red-bellied lemur (right) are 3 of the 42 different species that are commonly referred to as lemurs (which excludes the aye-aye, sifakas, and the indri). All of these prosimians are found only on the island of Madagascar.

Infraorder Loriformes
Family Loridae

The lorises and pottos, members of the family Loridae, have only a very short tail or one that is almost completely absent. They are nocturnal and arboreal, and all mark their territories with urine. Lorises and pottos move slowly through the trees and exert a viselike grip, which is used for confident quadrupedal movement as well as for securing themselves when sleeping. Unique among the primates, their hands and feet are adapted with dense arteriovenous networks, termed retia mirabilia (wonderful nets), which provide extremely good blood circulation and allow them to hold and grasp almost indefinitely without any loss of sensation.

All lorises are Asian. The appropriately named slender lorises (*Loris* spp.), with their thin bodies and stiltlike legs, are found in India and Sri Lanka. Reported to be aggressively territorial, slender lorises usually live alone or in association with a

Galagos, also called bushbabies, are able to move each of their ears independently to listen for sounds of prey or predators as they move through the forest.

mate. They are omnivores and hold on to branches with the feet while grabbing prey with their hands. The slow lorises (*Nycticebus* spp.) occur throughout the Near and Far East, including India, Bangladesh, Malaysia, Indonesia, Vietnam, and China. They are much bulkier by comparison but have a lifestyle and omnivorous diet very similar to the slender lorises.

The pottos (which include the angwantibos) are exclusively African and can be found in a wide band that crosses the central portion of the continent. They can be considered the African counterparts to the Asian lorises, occupying the same ecological niche. The pottos rely heavily on fruit in their diet and also consume the gum from trees as well as insects. They may also eat small amounts of meat. Pottos are reported to rely heavily on their sense of smell when foraging, and they move along with their nose against the branch.

Family Galagonidae

The galagos, also commonly called bushbabies, are members of the family Galagonidae, which includes 3 genera and 20 species. All are exclusively nocturnal and are found only in sub-Saharan African woodlands. Most species have an average body length of about 15 centimeters, a long tail, and legs that are much longer than their arms. They move primarily by leaping and can cover 2.5 meters in a single jump. Their digits have flattened disks of skin on the tips to assist in climbing and gripping. All galagos mark their territories with urine and engage in "urine washing." The hands and feet are anointed with urine during this behavior, which may function as a means for spreading the scent during locomotion as well as facilitating a more secure grip when climbing on smooth surfaces. Galagos live in small groupings that can include adults and offspring. As a family, they have a diverse diet of fruit, vegetation, tree gum, insects, and other animal prey such as small reptiles, birds, and mammals, but the different species may rely more on some of these components than on others. Some species rely heavily on tree gum and may visit 500 to 1,000 sites each night when foraging. Adults generally forage alone. The ears are large, and the earflap is mobile and expressive. It can be rotated, moved forward, backward, or even folded downward. One of the more distinctive behaviors of the galagos is that the infants are carried in the mouth, unlike most other primates. Adult galagos have a vocalization that closely resembles the cry of a human infant and may be the origin of "bushbaby," a commonly used name.

Suborder Haplorrhini

The haplorrhines are the "dry-nosed" primates. The species in this group have dry noses that are not fused with the upper lip. Since the lip is not attached to the gum (as it is in the strepsirrhines), haplorrhine faces have a much wider range of expressions. This group includes the Tarsiiformes (tarsiers) and the Simiiformes (monkeys and apes), which constitute the majority of species within the order Primate.

The haplorrhines are considered to have more advanced physiological and behavioral characteristics than the strepsirrhines. They have larger relative brain sizes compared with the prosimians, and their primary sense is vision, which is emphasized neurologically. Each eyeball is encased in a bony cup in the skull, and except for the nocturnal genera *Tarsius* and *Aotus*, haplorrhines are diurnal and have color vision. Faces generally lack the prominent snouts and ears that are common for strepsirrhines, and the senses of smell and hearing are deemphasized. Hands and feet are much more generalized in their appearance and usually have no specific anatomical specializations for grooming.

Physical and behavioral aspects of reproduction are also distinct. All of the haplorrhines (except for *Tarsius*) have a uterus with one chamber, unlike the more primitive two-chambered (bicornuate) uterus found in prosimians. Single births (rather than litters) are the norm, although twins and triplets are common for the

The haplorrhines, such as this olive baboon, are the "dry-nosed" primates. Their noses are not fused with the upper lip, and they are capable of a much wider range of facial expressions than strepsirrhines are.

marmoset and tamarin monkeys. All of the haplorrhines have a distinct type of placenta, described as hemochorial, in which the mother's blood is in direct contact with the developing fetus. Newborn infants of haplorrhines are relatively much larger than those of the strepsirrhines, despite similar gestation periods. Although all primates have an extended period of dependence on the mother, this is more obvious for haplorrhines and reflects the overall increase in complexity of their behavior and natural history.

Infraorder Tarsiiformes
Family Tarsiidae

All tarsiers are native to forests and bamboo groves of Southeast Asia and the Philippine Islands. They are primarily nocturnal and have a carnivorous diet that includes amphibians, reptiles, and insects. Reports of their social behavior suggest considerable flexibility. They have been seen traveling alone, in families, and in diverse groupings that may include several adults.

Tarsiers have a number of distinct characteristics. Their primary style of movement is leaping, and they may cover up to 5 to 6 meters in one jump. This is especially impressive considering that the total combined length of their head and body averages only 85 to 160 millimeters. The length of the legs is almost twice that of the body, and the name "tarsier" is derived from their elongated tarsal bones, which make up the ankle region. Their nearly hairless tail ranges from 135 to 275 millimeters in length. When tarsiers cling vertically, the tail is used as a brace to provide additional support. A "friction pad," present near the base of the tail, assists in this process. Each digit also has a flattened pad on the tip, which functions like a suction cup during climbing and especially when clinging.

Tarsier eyeballs are supremely adapted for collecting as much light as possible during their nocturnal hunts and are the dominant feature that characterizes this genus. They have the largest eyes relative to all other primates. Each eyeball is approximately 16 millimeters in diameter and is about the same size as their brain. Comparable eyeballs for humans would weigh about 1.4 kilograms each. In order to maximize their field of vision, they are able to rotate their head nearly 360 degrees.

For taxonomists, tarsiers have always been a source of keen interest and lively debate, since they do not fit neatly into either the strepsirrhine or haplorrhine category. Rather, they exhibit a blend of features that are found in both. Like the prosimians, tarsiers have a variety of anatomical features that are considered primitive. Two grooming claws are present on each foot. Multiple nipples and a two-chambered uterus are present, although single births are the norm rather than litters. Tarsiers regularly engage in prosimian style urine washing and mark their territories with urine as well. They also mark with secretions from scent glands, but this behavior is found in both prosimian and monkey species. The large ears are very mobile and can be moved independently.

In addition to these typical prosimian characteristics, tarsiers share many features with the monkeys and apes. The nose is dry and not fused with the upper lip. The lip is not attached to the upper gum, allowing the tarsiers to have a greater range of facial expressions than the prosimians. Their eyes do not have a tapetum lucidum cellulosum, and the eyeball is encased in a bony cup in the skull. Vision is the dominant sense, with a neurologically reduced emphasis on olfaction. In terms of reproduction, female tarsiers have monthly sexual cycles rather than a breeding season. Females also have a swelling of the skin around their genitals that is associated with ovulation, similar to many monkeys. The placenta is hemochorial, as is found in the monkeys and apes. Infant care illustrates the same intriguing blend of features that is typical of other characteristics of tarsiers. Young offspring cling to the mother and are carried in the usual style of monkeys and apes, but females also utilize the prosimian style and will pick up the baby and carry it with their mouth.

The monkeys are divided into two large groupings: the platyrrhines, or New World monkeys, and the catarrhines, or Old World monkeys. Howler monkeys (left), native to Mexico, Guatemala, and Belize, have nostrils with the characteristic outward flare found in the platyrrhines. The olive baboons (right), found in many parts of Africa, have nostrils that point straight down, like all catarrhines.

Tarsiers have features that straddle both the strepsirrhine and haplorrhine boundaries, which has been interpreted differently by various taxonomists and primatologists. On the basis of tarsiers' unique blend of features, some sources consider them to be prosimians, while others do not. Currently, the argument in favor of designating them as haplorrhines dominates the field and will be followed for this book.

Infraorder Simiiformes—The Monkeys and the Apes

The monkeys are divided into two sections, the Platyrrhini and the Catarrhini. Each of these has both a geographical and an anatomical association. The platyrrhines are the New World monkeys, occurring only in the southernmost portion of North America and in Central and South America. The name of this group is derived from the Greek words *platy* (flat or broad) and *rhin* (nose). These primates have "broad" noses, meaning that the nostrils flare outward and are widely spaced. *Catarrhine* is derived from the Greek words *kata* (down) and *rhin* (nose). The "down-nosed" primates have nostrils that point straight down and are spaced closely together. Officially, humans and the other apes fall into the catarrhine category, but the term is generally used only in regard to monkeys. All of the catarrhine species are considered Old World monkeys, found throughout Africa, Asia, and a small portion of the Arabian Peninsula. The exact shape and position of the nostrils within each of these infraorders do vary, and there is no absolute standard for comparison. While the trend in each infraorder is clear, the terms *platyrrhine* and *catarrhine* are used most reliably in reference to geographical origin rather than physical appearance.

Platyrrhines—The New World Monkeys

The species in this group are diverse in both their body types and natural histories, but all have primarily arboreal lifestyles. Unlike many of the Old World monkeys, female New World monkeys have no external swelling of the area around their genitals, which is associated with ovulation. An additional distinction is that none of the New World monkeys have cheek pouches for holding food. The platyrrhines are divided into four families, and each is composed of several smaller groupings of similar species. The following subsections summarize the relationships among all of the different types of New World monkey.

FAMILY CEBIDAE

The marmosets and tamarins are the smallest of the New World monkeys. Adult pygmy marmosets, *Callithrix pygmaea*, may weigh only 100 grams and are the smallest of any monkey species. Newborn pygmy marmosets are only about the size of a lima bean. Aside from their relatively small size, marmosets and tamarins share a large number of physical and behavioral features. They have claws on all of their digits except for the big toe, and their thumbs are not opposable. Males and females are the same size. In some species, such as the golden lion tamarin (*Leontopithecus rosalia*), males and females establish a monogamous pair bond that forms the core of the social unit along with one or two generations of their offspring. The adult male, as well as older siblings, directly assists in the care of the young. Other species, such as the cottontop tamarin, *Saguinus oedipus*, may have less stable social groupings.

Marmosets and tamarins occupy similar habitats, usually preferring the midlevel of secondary forest. Both eat a variety of foods, including fruits, invertebrates, small

Squirrel monkeys are found throughout Central and South America, sometimes forming temporary groupings that number in the hundreds. Males have a significant increase in weight and aggressive behavior during mating season, and they compete for dominance. Females prefer to breed with higher-ranking males.

vertebrates, and tree sap. Marmosets rely much more heavily on sap consumption than tamarins do and have dental adaptations associated with this behavior. All species of marmoset and tamarin (except for Goeldi's marmosets) have two incisors, one canine, three premolars, and two molars in each half of both jaws. The lower canine teeth of marmosets are shorter and blunter than those of tamarins. These teeth are used to bore into tree trunks to release sap in much the same way as humans use taps to collect maple tree sap for syrup. The longer and comparatively thinner canines of the tamarins are never used for boring into wood. The importance of this food source for marmosets, as well as the dental adaptation, is illustrated by the fact that pygmy marmosets may spend over 70 percent of their foraging time acquiring and consuming sap. Tamarins will occasionally eat gum that they encounter but do not actively gouge at trees to initiate the flow of sap.

The squirrel monkeys (*Saimiri* spp.) and the capuchin monkeys (*Cebus* spp.) are the remaining members of Cebidae. These two genera of primates have the most varied diets of any New World monkey, routinely consuming a large range of items such as fruits, insects, and nuts. Squirrel monkeys are particularly dependent on insect prey, which makes up a substantial portion of their diet. When foraging for food, capuchins display remarkable energy and curiosity as they move through the forest. With the combination of opposable thumbs and keen mental abilities, capuchins are universally recognized as the most dexterous of the New World monkeys. Their hectic style of search and investigation has earned them the title of "destructive foragers."

Both groups are commonly seen in the same habitat, although capuchins appear to prefer the canopy while squirrel monkeys usually inhabit lower levels of riverine forest. *Cebus* live in fairly large social groups with multiple adult males, females, and offspring. Determining exact sex ratios in a group can be tricky, since female genitals look very much like typical male anatomy. *Saimiri* are known to form extremely large groups, which have been estimated to have as many as 500 individuals. More commonly, their groups average from 20 to 50 individuals of both sexes. Female squirrel monkeys are usually dominant to males, except during breeding season. At that time, males have a temporary increase in weight and become more aggressive. This change in size and behavior is related to mating success. Larger males are able to intimidate smaller males, earning them higher status, or dominance, in the group and a greater opportunity to reproduce because of the resulting preference that females have for them. All females have single births, and because squirrel monkeys have a breeding season, the births are clustered in a short period once a year. Males have no direct role in caring for offspring; however, this species exhibits "aunting" behavior, in which infants receive care from multiple females in addition to their mother. Unlike squirrel monkeys, capuchins do not have a breeding season. Typical for most of the other New World monkeys, females have predictable cycles, single births, and no aunting behavior. Capuchins are exceptional in that they are the only species outside of Atelidae that have prehensile tails. Unlike the members of Atelidae, their tails have a comparatively limited function and are not used for locomotion but rather are fully covered with hair and used primarily for support during foraging.

FAMILY NYCTIPITHECIDAE

The night monkeys (*Aotus* spp.) are also commonly referred to as owl monkeys or douracoulis. They are the only nocturnal monkey. Characteristically for animals that spend their waking hours in the dark, their eyes are very big in order to collect as much light as possible, and they do not have color vision. Unlike the prosimians, their eyes do not have a tapetum lucidum cellulosum, which reflects light. During the day, they sleep in tree hollows or tangles of vegetation in branches. They become active at night and forage primarily for fruit, although they will consume a variety of other foods as well. They appear to utilize all levels of the forest and occupy a range of habitats.

The night monkeys are the only species of nocturnal monkey. These night-dwellers do not have color vision, unlike most other primates.

As is typically the case for nocturnal species with many potential predators, night monkeys live in very small groups, which are reported not to exceed five individuals. The core of the group is a monogamous pair, and the rest are their offspring. Both of the adults actively care for their offspring. Juveniles are primarily carried by the male and are returned to the mother only for nursing.

FAMILY PITHECIIDAE

Sakis (*Chiropotes* spp. and *Pithecia* spp.) have been reported to prefer the midlevel of the forest. They are normally quadrupedal and make impressive leaps between trees. In addition, sakis are able to run along branches bipedally while keeping their arms extended upward. They live in groups that may have 20 or so members and communicate with each other using very loud calls. Their teeth have adaptations that allow them to exploit difficult foods. The forward tilting incisors and short, strong canines are used to open fruits and nuts that have very hard outer surfaces. The softer interior provides a nutritious meal. Flattened molars function like a mortar and pestle for grinding and processing hard seeds. Unrelated to their unique dental features, sakis have been documented to eat small birds and bats that they extract from tree hollows.

The uakaris (*Cacajao* spp.) are similar to the sakis in their lifestyle and have the same feeding specializations. They occupy the canopy level of the forest and live in groups that may include as many as 100 individuals but more commonly have between 10 and 30 members. Uakaris are best known for their distinctive look. Hair color varies by subspecies, ranging between white, orange, and chestnut. Most notable is the striking appearance of *C. calvus*, whose head is completely bald and skin is bright red.

Titi monkeys (*Callicebus* spp.) are distinct from the sakis and uakaris. They inhabit the lower areas of the forest and live in small groups based on a monogamous pair. Both the adult male and adult female take active roles in the care of their offspring. Titis have a delicate appearance and are very agile as they move throughout the forest in a quadrupedal style. Frequent vocalizations are common among the members of the family group. These monkeys exert little control over their very long, thin tails. However, when sitting closely to each other, they commonly intertwine their tails into one long braid, giving the appearance of complete contentment.

FAMILY ATELIDAE

The species within Atelidae are very similar in appearance and general behavior. All prefer the canopy level of primary forest, live in extended social groups, reproduce throughout the year, and generally have single births. The diets of the atelids do have some variation among species. Spider monkeys (*Ateles* spp.), for example, are primarily frugivores (fruit eaters), while the muriquis (*Brachyteles* spp.) are much more folivorous (leaf-eating).

The universal characteristic for all members of Atelidae is the presence of a prehensile tail that is capable of a variety of functions. It can be used for grasping, holding, climbing, swinging, reaching, and numerous other behaviors. In contrast to the prehensile tails of capuchin monkeys (*Cebus* spp.), which are much more limited in their range of ability, the tails of the atelids provide the equivalent of a third hand or foot. Their tails are constantly in use but particularly important during feeding and locomotion. All of the atelids are capable of suspending themselves solely by their tails, freeing their hands and feet for foraging and feeding. Spider monkeys (*Ateles* spp.) have the most dexterous of the prehensile tails and use them interchangeably with their hands during their "semibrachiation" form of locomotion. An additional adaptation that facilitates suspended locomotion for some species in Atelidae is the drastic reduction, or complete absence, of thumbs. The hands function more like hooks, which allows for very efficient and rapid movement through the intricate pathways in the forest canopy. Both the spider monkeys

This spider monkey exhibits the long limbs and prehensile tail that characterize all of the atelids.

and woolly monkeys (*Lagothrix* spp. and *Oreonax flavicauda*) are also known to move or stand bipedally on the ground or on the surfaces of large branches. Spider monkeys simply hold their tails out of the way, while woolly monkeys may use them to form a tripod as they stand.

Catarrhines—The Old World Monkeys
The Old World monkeys are categorized as one family, Cercopithecidae, which has 131 species. The primates in this group occur throughout Africa, Asia, and a small portion of the Arabian Peninsula. This large family is divided into two subfamilies primarily on the basis of dietary habits and related physical features. The species that are principally leaf eaters are the Colobinae, and those that have a much more varied diet are the Cercopithecinae. The cercopithecines all have simple stomachs and cheek pouches for storing food as well as arms and legs that are roughly equivalent in length. They have prominent ischial callosities, which are leathery areas of skin on their rear ends that function primarily as sitting pads. The colobines do not have cheek pouches, and their ischial callosities are either greatly reduced or completely absent. Their legs are generally longer than their arms, related to their style

A female spider monkey uses "semibrachiation" to move through the canopy. Her prehensile tail reaches and grasps branches interchangeably with her hands.

of locomotion, which relies heavily on jumping and leaping. All colobines have chambered stomachs, a specialization that allows them to efficiently digest a diet that is composed primarily of leaves. Both the cercopithecines and the colobines have the same dental formula of two incisors, one canine, two premolars, and three molars (2.1.2.3). However, the shape of the teeth in each subfamily corresponds to their dietary differences. The folivorous colobines have higher, more pointed cusps on their molars, while those of the omnivorous cercopithecines are comparatively flat. No Old World monkeys have prehensile tails.

SUBFAMILY CERCOPITHECINAE

Swamp Monkeys   The single species of swamp monkey (*Allenopithecus nigroviridis*) has some features in common with both the guenons and the baboons. While visibly similar to the guenons in overall appearance, swamp monkeys are distinct in that females have swelling of the skin around their genitals that is coincident with ovulation. This is unknown for the guenons but common for baboons. The sitting pads of the males are fused, unlike most other species, where males have two parallel pads. Swamp monkeys consume a variety of foods, including fruits and insects. They are excellent swimmers and divers and are reported to supplement their diet with fish.

Talapoins   Talapoins (*Miopithecus* spp.) are the smallest of all the Old World monkeys, with adult males weighing about 1,200 grams and adult females averaging around 800 grams. They resemble the squirrel monkeys (*Saimiri* spp.) and occupy a similar ecological niche. The talapoins live in very large groups that may have around 100 members, with multiple adult males and females. Their preferred habitat is swampy or mangrove forest, and they are never far from water. Talapoins are excellent swimmers and are equipped with webbing between their fingers and toes. Their diet is diverse and includes insects, fruits, leaves, eggs, and even shrimp.

Patas   Patas monkeys are the lone species in the genus *Erythrocebus*. These savanna-dwelling primates are described in detail in *Which Primates Are the Fastest?*

Vervets   The monkeys that compose the vervets (*Chlorocebus* spp.) are widespread throughout Africa. Primary habitat is woodland savanna, but vervets are very adaptable and may live in diverse areas such as mangrove swamps or near human populations. All of the six species in this group are equally comfortable on the ground or in the trees and live in fairly large groups that contain multiple adult males and females. Vervets are true omnivores, with a diet that includes fruit, in-

Patas monkeys are found in African grasslands and savanna and are the only species in the genus *Erythrocebus*.

sects, eggs, and even small vertebrates. Those that live in mangrove forest spend large amounts of time catching and eating fiddler crabs.

The vocalizations of vervets have been well studied. These monkeys utilize a number of different calls to identify specific types of predator. Infants and juveniles learn how to produce and use these referential calls over time, becoming more competent as they mature. Vervets are one of the few species, other than humans, that demonstrate the ability to label things in their environment with a vocalization (see *Do All Primates Have Their Own Languages?*). Future studies with other species will be necessary to determine how common this skill is within the primates.

A vervet monkey receives grooming from another group member.

Guenons The guenons (*Cercopithecus* spp.), an African group, are composed of 25 different species, making them the most common type of monkey on the continent. Forest is the preferred habitat for guenons, and all are primarily arboreal, although perfectly comfortable on the ground as well. They use their very long tails for balance during leaping and climbing, although infants demonstrate limited prehensility that disappears as they mature. None of the guenons show any sexual swelling associated with ovulation. There is no defined breeding season, but births are frequently clustered during the time of year when food is more plentiful. Most species tend to live in relatively small groups with one dominant male. However, because of the large number of species present in this group, accurate generalizations about social structure and behavior are difficult.

The most distinctive feature about guenons is the remarkable diversity in their coloration and patterning. Each species has a very distinctive appearance that can be used for identification, and guenon faces have been compared to those of kabuki actors in full stage makeup. The De Brazza's guenon (*C. neglectus*) is a remarkable example. The faces of adult males have a black cap, an orange crescent on the forehead, a black mask, a powder blue patch over the nose, and a dramatic white beard. The diana guenon (*C. diana*) is less colorful but has bold, contrasting patterns of light and dark colors and lines over its body. In addition to providing an easy means of species identification, the colors and patterns on guenons are clearly adaptive as well. The two most likely explanations for these variations are related to their forest habitat. While primates see in color, their natural predators usually do not. Instead of making the guenons more obvious, contrasting colors and patterns most likely mimic the splotches of light and shadow throughout the forest and allow the monkeys to blend effectively with the background. When the monkeys communicate with each other, strong colors and lines most likely emphasize body movements and expressions. In the forest environment, the body gestures that are used among the monkeys can be "read" more easily and make interactions more effective.

Macaques   The macaques (*Macaca* spp.) are the most adaptable and geographically widespread of all the primates, with the sole exception of humans. They occur in northern Africa and throughout Asia including India, China, Thailand, and as far to the east as Japan. Populations of free-ranging macaques have been established outside their traditional ranges in locations such as Puerto Rico, France, and Florida. Even western Europe has a small population of introduced Barbary macaques (*M. sylvanus*) that have lived on Gibraltar since Roman times. All macaques are perfectly at ease on the ground or in the trees, and a number of species are accomplished swimmers and divers. Their habitats range from tropical forests to snow-covered mountains as well as urban areas that are densely populated with humans.

All macaques are similar in body type, although average size can vary considerably among the 20 different species. Also, physical features, such as hair and skin color or presence of a tail, vary among species. Some macaques are seasonal breeders, while others have predictable cycles. All give birth to singletons, and females are usually the sole providers of infant care. The clearest exceptions are adult male Barbary macaques, which take an active part in the care of youngsters.

The overwhelming similarity among macaques is their adaptability, a unifying characteristic of this group. All macaques are extremely dexterous, demonstrating levels of opposability and precision comparable to that shown by humans. Their remarkable manipulative abilities, combined with their active minds and inquisitive natures, have allowed them to exploit a range of habitats unknown to most other primates. Except for humans, the macaques are the most successful group of primates.

Mangabeys   The mangabeys include two genera (*Lophocebus* and *Cercocebus*) and a total of nine species. The members of *Lophocebus* are most closely related to the

Lion-tailed macaques, native to India, are the most arboreal of all the macaques.

baboons (*Papio* spp.), and those of *Cercocebus* to the mandrills and drills (*Mandrillus* spp.). Despite their distinct lineages, the mangabeys share physical and behavioral similarities. All are large, slender monkeys that inhabit the rain forests of Central Africa. Mangabeys spend most of their time in the trees, although they will also descend to the ground (especially *Cercocebus*) in search of food. There are primarily frugivorous but supplement their diets with seeds, nuts, vegetation, and invertebrates. Mangabeys have obvious cheek pouches that they fill with prodigious amounts of food as they forage. All have a variety of loud and distinctive vocalizations. Ischial callosities are well developed, and in *Cercocebus*, those of the male are fused across the middle. Females have swellings of the sexual skin that is associated

Like all primates, olive baboons are highly tolerant of youngsters. Here an adult male gently interacts with an infant as its mother looks on.

with ovulation. Social organization is variable but generally includes groups that range from one dozen to two dozen members.

Baboons, Mandrills, Drills, and Geladas It is common to find baboons, mandrills, drills, and geladas all under the general label "baboons." However, this is misleading because taxonomists now divide these monkeys into three genera. Only the five species in *Papio* are correctly referred to as baboons. Mandrills and drills are each a different species of *Mandrillus*, and the geladas are the only species in *Theropithecus*. The members of these three genera are the largest of all the monkeys, exhibiting extreme sexual dimorphism, in which adult males may be twice the size of adult females. For example, male *Papio hamadryas* may weigh around 27 kilograms, while females are closer to 13 kilograms.

The baboons were known in ancient times, commonly represented in the writings and sculpture associated with Egyptian mythology. The faces of these monkeys have a doglike appearance, and males have bodies that are particularly bulky and muscular. All baboons are primarily terrestrial but will ascend into trees at night for sleeping. They occupy habitats that range from open savanna to wooded areas. Their diets are very diverse, including all forms of vegetation as well as invertebrates. They actively hunt for meat and regularly consume small mammals such as hares.

Baboon social organization is unique among the primates. Generally, one adult male aggressively defends a small number of females, forming a small group. These groups may travel and forage on their own or congregate with other similar groups. In *Papio hamadryas*, these congregations can be quite large, possibly containing 100 individuals that forage simultaneously. Associations as large as 750 individuals have

Olive baboons live in a complex social system. Large groups are composed of smaller subgroups that contain one adult male, multiple adult females, and their offspring.

been documented to occur in sleeping sites. These groupings are flexible and can change in response to environmental conditions. When food is scarce, groups space themselves widely as they forage. If sleeping sites are limited, large numbers of these baboons will tolerate each other in a small space. Males without any females are also commonly seen. Grooming between baboons is a very regular behavior and is most often given to males by females. All baboons have large ischial callosities, and females have very prominent sexual swellings associated with ovulation.

Mandrills and drills are closely related to the members of *Papio* and share some of the same characteristics. Like the baboons, these two species of *Mandrillus* show extreme sexual dimorphism. They also live in small, male-dominated groups that sometimes congregate with other groups to form associations of up to 100 individuals. Unlike the baboons, mandrills and drills are found only in forested areas. Mandrills are best known for their stunning facial coloration. While all adults have colorful faces, the skin of the adult males is particularly vibrant. Framed by dark hairs that appear tipped with gold, adult male faces have a neon red stripe that runs parallel from the brow ridge and covers the nostrils. Each side of the nose has a broad band of sky blue mixed with tinges of purple. The lips are pink, and the beard is bright orange. As with the guenons, the bright colors are most likely an adaptation that facilitates communication and group cohesion in the shadowy areas of the forest. Adult males also make themselves obvious to those that are following them, since their rear ends are as colorful as their faces. Mandrills are the monkeys with a rainbow in front and a sunset behind. Drills are distinct, having a handsomely dark face but sharing the dramatically colored posteriors.

Geladas   Geladas are the only species in their genus, *Theropithecus*, and are closely related to the members of *Cercocebus*. Native to the mountains of Ethiopia, these terrestrial monkeys subsist almost entirely by foraging on grasses. This unique lifestyle makes them the only primate that can be considered a grazer, with up to 90–95 percent of their diet composed of grass. The abundance of food and consequent lack of competition allow hundreds of these monkeys to coexist in a relatively small area. They are also unique in that they forage in a sitting position rather than in the quadrupedal fashion of most other primates. They exhibit extreme sexual dimorphism, and both males and females have a bright red hourglass shape on their chests. The color of this area deepens when females are ovulating, and slight swellings emerge that resemble the shape of a necklace. These visual signals communicate a state of sexual readiness to males. Threats, especially between males, are communicated by moving the scalp backward, which exposes the light colored eyelids. Subordinate males have a dramatic means of showing their deference by flipping their upper lip backward, completely exposing their gums and teeth.

SUBFAMILY COLOBINAE

Colobus   The colobus monkeys are the African foliovores. This group includes 3 genera and 15 species, and each genus is usually identified by color. *Colobus* are black and white, except for *C. satanas*, which is completely black. *Piliocolobus* all have some areas of red hair, and the lone species in *Procolobus* is an olive color. The colobus monkeys have several features in common. All are primarily arboreal and have thumbs that are either greatly reduced or completely absent. The name of this group is derived from the Greek word *kolobos*, meaning "docked" or "mutilated," in reference to this natural characteristic.

This anatomical specialization facilitates movement through the trees by using the hands and arms like hooks, similar to the New World atelids. Colobus monkeys all have sacculated stomachs that are populated with bacteria that can break down the cellulose in their leafy diets. Ischial callosities are present in all species, and cheek pouches are absent. Each of the three genera in this group have distinctive characteristics.

The five species of *Colobus* are spectacular in their appearance. Most have tails with a luxuriant white tuft of hair on the end and long white hair in a capelike pattern on their backs. These features are emphasized as they leap throughout the forest canopy. They are usually found in small groups with one adult male and several females with offspring, but they may also congregate in bands that have been reported to number as many as 300 individuals. Home ranges are considered fairly small, clearly influenced by the abundance of edible vegetation that can be found with limited effort. Their three-chambered stomach allows them to consume a diet

Black-and-white colobus are one of the most distinctive and beautiful species of Old World monkey. They live in the canopy of the Africa forest and subsist primarily on vegetation. Adult males advertise their presence with roars that can be heard at great distances.

that is predominantly composed of leaves, but they supplement their diet with flowers, some fruits, and other assorted types of vegetation. *C. satanas* is distinct, favoring a diet that is rich in seeds rather than leaves.

Black-and-white colobus reproduction is not limited to a breeding season, and females show no sexual swellings related to ovulation. Infants are born perfectly white, gradually acquiring the typical black and white pattern over the first few months of life. All females will readily carry and provide care to any infant in the group, although the offspring are quickly given to their mother for nursing. Males provide no direct care for infants but will defend the group against predators or rival males.

The red colobus, in the genus *Piliocolobus*, include nine species. These monkeys are slightly smaller than the members of *Colobus* and live in larger groups with adults of both sexes. Red colobus have a four-chambered stomach and more diversity in their diet than the black-and-white colobus. An additional distinction is that females have obvious swellings associated with ovulation, similar to the cercopithecines. Red colobus are also particularly well known as favored prey for chimpanzees (*Pan troglodytes*). In some areas where their ranges overlap, the red colobus population may be depleted by 10 percent annually because of chimpanzee hunting parties.

The olive colobus are the least studied of all the colobus monkeys. This single species in the genus *Procolobus* is the smallest of the colobines and often occurs in the lower levels of the forest as well as on the ground. As in the red colobus, females have obvious sexual swellings associated with ovulation. Unlike any other monkey or the apes, olive colobus females are known to carry their offspring with their mouth in a style that is reminiscent of prosimian behavior.

Langurs This large group includes all of the colobines that are commonly referred to as langurs and leaf monkeys, spanning 3 genera (*Presbytis*, *Semnopithecus*, and *Trachypithecus*) and 35 species. All are exclusively Asian and are much more diverse than their ecological counterparts in Africa. The members of this group all have thumbs, long limbs, and slender bodies. They move quadrupedally or make impressive leaps between treetops. Both coat color and facial coloration vary considerably among the species, ranging from the uniformly dark faces and silvery gray hair of the sacred langurs (*Semnopithecus* spp.) to the nearly brick red color of the maroon leaf monkey (*Presbytis rubicunda*), which has a bluish gray face with pinkish lips and chin. While all infant leaf monkeys are lighter than adults, the neonatal coloration exhibited by some species is striking. For example, Phayre's leaf monkeys (*Trachypithecus phayrei*) are brownish as adults, but infants are born ranging from yellow to a reddish orange. This distinctive coloring slowly disappears, and adult coloration emerges as the infant matures. As in the African colobines, distinctive neonatal coloring adds to the special appeal of infants, and aunting behavior is very common for all members of the langur group. A notable distinction for this group is the social structure of the Mentawai langur (*Presbytis potenziani*), reported to be the only monogamous species of Old World monkey.

The leaf monkeys are all diurnal and arboreal, and their diet is primarily composed of vegetation. They occur throughout Asia, with their range including Borneo, China, Vietnam, and Laos. The langurs are more terrestrial than the leaf monkeys and have an omnivorous diet. The sacred langurs are the most widespread and probably most familiar genus in this group. They are extremely adaptable and can be found in a wide range of habitats that include tropical, deciduous, and coniferous forests, very dry areas near desert, and throughout the Himalayas. Sacred in the Hindu religion, these langurs are also common in urban areas and have the ability to coexist with humans.

An additional measure of the behavioral flexibility of sacred langurs is reflected

The silvery leaf monkey can be found in Borneo, Sumatra, and smaller islands in Southeast Asia. The dramatic coloration of this infant will slowly fade during maturation, eventually matching the adult form.

in the variation that exists in their social structure. Troops may have fewer than 10 or as many as 100 individuals. Each group has one dominant male and a stable core of adult females. Groups of bachelor males exist as well. Dominant males in mixed sex groups are not permanent and are usually deposed by a group of bachelor males. During an aggressive encounter when a dominant male is removed, infanticide by the new male is common. This reproductive strategy removes the rival's offspring and hastens the female's next reproductive cycle. While far from universal for all primates, this harsh series of behaviors perfectly illustrates the basic tenets of the Darwinian theory of evolution.

Odd-Nosed Primates This final grouping of Old World colobines can be fairly described as an informal collection of species with very distinctive noses that are not typical for any of the other primates. Behaviorally and ecologically, they are all very similar to the other colobines with the exception of their remarkable profiles. This group includes four genera and nine species.

The colorful doucs (*Pygathrix* spp.), snub-nosed monkeys (*Rhinopithecus* spp.) of China, and the pig-tailed langur (*Simias concolor*) all have small, delicate looking noses that appear turned up at the tip. The last species in this group is quite differ-

The exaggerated noses of adult male proboscis monkeys are the evolutionary result of female mate choice. Males with bigger noses are preferred, leading to male offspring that carry the genes for this dramatic appendage.

ent. Proboscis monkeys (*Nasalis larvatus*) are found only in Borneo and inhabit swampy areas that border rivers. Their diet is composed chiefly of leaves, and they never venture far from the water as they forage. If threatened, these monkeys make spectacular leaps out of the trees and splash into the river, confidently swimming to safety. Their most distinctive feature, however, is their dramatic appearance. Adult males have enormous noses that hang over their mouths and, in some cases, literally have to be held out of the way in order to eat. Females are spared this appendage, having unremarkable noses that resemble those of the pig-tailed langur. Clearly then, the large proboscis of the males is unnecessary for survival and may even be a burden. The most likely explanation for this disparity in appearance is a classic example of mate choice by females. Like the elaborate tail of a peacock, the large nose of proboscis males signifies health and vigor to females. A male's nose advertises that, even with the possible impediment, he has survived and flourished. For females, a large nose on a successful male is a reliable indication of desirable genes for her offspring. Stated directly, the bigger the nose, the sexier the male. Beauty, as always, resides in the eye of the beholder.

The apes, for the majority of people, are probably the most familiar and easily recognized members of the primate order. Although well known, the apes are the least successful type of primate, having a very small number of species compared with all other groups. The ancestors of modern apes diverged from the Old World monkeys approximately 25 million years ago. Except for modern humans (*Homo sapiens*), all species of ape are primarily restricted to the tropical forests of Africa and Southeast Asia.

Anatomically, the apes share a variety of features. External tails are completely absent, as are cheek pouches. The ape dental formula of two incisors, one canine, two premolars, and three molars (2.1.2.3) is identical to that of the Old World monkeys, but the general morphology of the teeth is distinct. For example, apes lack the dramatic sexual dimorphism in canine size that is standard for many of the cercopithecines. All of the apes have a degree of shoulder rotation that greatly exceeds what is possible for the majority of the other primates, allowing locomotion with only the arms. This ability is discussed in more detail under the question *How Do Primates Move?*

FAMILY HYLOBATIDAE

The members of this group, known commonly as gibbons, are referred to as lesser apes. This label denotes the relatively small size of the gibbons compared with the great apes. The hylobatids have greater diversity and a larger number of species than the great apes. All have ischial callosities, and females have minor swelling that is associated with ovulation. Gibbons occur in the forest canopy and subsist primarily, but not exclusively, on fruit.

These apes have a highly specialized form of suspensory locomotion known as brachiation and are the only true brachiators. Directly associated with this skill are their arms, which are the longest of all the primates relative to body size. Brachiation is discussed in detail under the question *How Do Primates Move?* The social system of the hylobatids is also distinctive. These apes live only in monogamous pairings, along with their recent offspring, and aggressively defend their territories. Pairs advertise their presence with elaborate and prolonged vocalizations that travel for great distances throughout the forest canopy. These songs vary by species, and most contain a portion that is produced as a duet between male and female. The singing of these songs reinforces the bond between the pair and also serves to repel other gibbons that may be nearby.

Direct care of offspring by both the adult male and female is common in gibbon families. The northern white-cheeked gibbon (*Hylobates leucogenys*) provides an example where physical appearance and parenting behavior have evolved to complement each other. These gibbons exhibit sexual dichromatism. Adult females are

White-handed gibbons, like all members of the Hylobatidae, have extremely long arms that facilitate their unique form of suspensory locomotion. Unlike all other apes except humans, gibbons live in monogamous pairings.

blond, with a black patch of hair on the top of their head. Adult males are completely black except for bands of white hair on the sides of their face, or "white cheeks." Both male and female infants are born completely blond and are cared for exclusively by the mother during the first 12 to 18 months of life, blending perfectly against her hair. Around that time, all infants gradually turn black, except for their white cheeks. At this point, they begin to be carried exclusively by the father, blending perfectly against his hair. The juveniles spend most of their time with the adult male until adolescence. The appearance of the male offspring remains unchanged into adulthood, but adolescent females undergo a third transformation. These individuals gradually become blond again but retain a black cap on the top of their head that is a remnant of their juvenile color stage. Now old enough to reproduce, they will be exactly the same color as the newborn infants that they will carry.

The gibbons are one of the most anatomically and behaviorally specialized of the primates. They are perfectly adapted to life high in the treetops and are the absolute masters of movement in their environment. While their unique blend of characteristics allows them to flourish in the forest canopy, they are also immensely susceptible to forest degradation and destruction. Unlike many of the other primates, gibbons do not have the ability to adapt and thrive in a range of environmental conditions.

Three members of the family Hominidae, gorillas (left), chimpanzees (center), and bonobos (right), are native to Africa and among the most cognitively complex primates. Orangutans and humans are the only other members of the Hominidae.

FAMILY HOMINIDAE

There are four genera and a total of seven species included in this family. The gorillas (*Gorilla* spp.), the orangutans (*Pongo* spp.), and the chimpanzee and bonobo (*Pan* spp.) are commonly referred to as the great apes. Modern humans (*Homo sapiens*) are the remaining species in the family. Although humans are taxonomically grouped among the great apes, they have distinct characteristics, which will be discussed separately here. The great apes have arms that are longer than their legs and usually locomote using all four limbs simultaneously. Humans have legs longer than their arms and are the only species of primate known to be exclusively bipedal as adults. All of the hominids have short trunks and hands that are similar, although the degree of opposability of the thumb varies by species. The great apes all have opposable big toes, which are used for grasping and climbing, while humans do not. All species have hair that is distributed uniformly over the body, although humans exhibit the most variation in hair color, length, and density of follicles. Humans and orangutans are sexually dimorphic regarding the appearance of the hair; males of both species generally have much more obvious hair growth over the body and the face.

The orangutan is the only Asian great ape, restricted to the islands of Borneo and Sumatra. Gorillas, chimpanzees, and bonobos are exclusively African and occur in a band that stretches across the middle of the continent. The ancestors of all modern humans were African but migrated thousands of years ago to eventually inhabit (or at least visit) all of the land masses on the planet.

Each species of hominid has a distinct natural history as well as a unique social organization. Females all have similar gestation lengths, and offspring depend on care given by adults for many years prior to self-sufficiency. Humans are the only species in which males directly care for infants.

The hominids are the most cognitively complex of all the primates. Unlike any of the prosimians or monkeys, great apes and humans learn to recognize their reflection in a mirror. All hominid species are known to use and make tools and are capable of learning abstract concepts such as the basic elements of language. Current research is revealing the extent to which the great apes are able to understand a situation from another individual's point of view and perhaps be deceptive for personal gain. While there is no doubt that adult humans have mental abilities distinct from those of the great apes, most likely the distinction resides simply in degree, not in kind. Presently, humans are the only species of hominid that is not in severe decline and facing extinction within a generation.

WHAT MAKES A PRIMATE A PRIMATE?

A number of different orders make up the group of animals known as the mammals. Frequently, all of the members of an order share a specific physical feature that distinguishes them from all other orders. The members of Rodentia, for example, all have incisors that grow continuously during their lives. Bats (order Chiroptera) are the only mammals that have wings, and any animal with an even number of digits on each foot, such as a sheep, belongs in Artiodactyla. Unlike members of other orders, primates do not have a single characteristic that defines them. Rather, they share a variety of physical features that make them unique.

Bodies

Although different species of primate vary markedly in their size, weight, and lifestyle, all primate skeletons are fairly similar in their general design. Unlike some mammals (such as cats), all primates have a clavicle, commonly referred to as a collarbone. The clavicle increases the overall mobility of the shoulder joint, and the range of mobility that is present is one way to distinguish the different types of primate.

All primates have a cecum that is located at the junction of the small intestine and the colon. This pouchlike organ houses bacteria that are essential for breaking down cellulose, which is the main component of the cell walls of plants. In humans, the cecum is poorly developed and considered a vestigial organ (the appendix protrudes from the cecum). As a result, humans are unable to digest many of the plant products that other species of primate regularly consume.

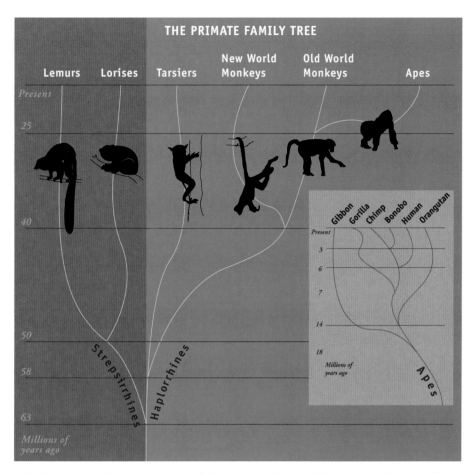

This chart presents the major groups and their estimated points of divergence within the order Primate. The inset details the evolutionary history of the species that make up the apes. While both the age and relative complexity of each radiation may vary, no species should be considered "more evolved" or "less evolved" than another. Groups that emerged earlier than others may be termed "primitive" in comparison. All are adapted to their particular environments and have survived by outcompeting other species.

Reproduction

Primates generally have two breasts, which are located on the chest, not the abdomen. Male primates' penis and scrotum are pendulous rather than being attached on the lower abdomen or inside the abdomen.

Teeth

Primate teeth are adapted for a generalized diet, and individuals have two full sets during their lifetime. The first set erupts very early in life and is deciduous, being replaced later in life by a permanent set that may last until death. Primates have four types of teeth: incisors, canines, premolars, and molars. The premolars and mo-

The anatomy of all primates except humans allows them to consume and digest very tough forms of vegetation.

lars are also known as the cheek teeth and have low crowns on their chewing surface. The upper jaw (maxilla) and lower jaw (mandible) each have bilaterally symmetrical halves, and a dental formula is derived from the number and type of teeth present in each half. Most primates have a dental formula of 2.1.2.3, which is translated as two incisors, one canine, two premolars, and three molars in each half of the jaw. The incisors and canines, along with the lips and tongue, are generally used for moving food into the mouth. Premolars and molars are used for chewing the food, which is the first step in digestion.

Hands and Feet

The hands and feet of all primates are adapted for efficient grasping and climbing. Most species exhibit pendactyly, meaning there are five fingers on each hand and five toes on each foot. All primates have fingerprints and toeprints, which can be used to identify individuals. In general, flat fingernails or toenails have replaced claws. These nails are used actively in a large range of activities, such as scraping, opening, cleaning, and scratching. In addition, they also have passive roles. Each nail functions somewhat like a turtle's shell; the hard surface and curved shape pro-

This orangutan illustrates the dental formula shared by most species of primate: two incisors, one canine, two premolars, and three molars in each half of the jaw.

tect the sensitive tip of the digit beneath. Also, the rigid structure acts as a solid backing to support the ends of the fingers or toes during their active uses in gripping, climbing, and manipulating objects.

One of the most distinctive features that characterize primates is the presence of a big toe and/or a thumb that has greatly enhanced mobility. All species in this order (except humans) have a big toe that is divergent from the other toes and can be used for grasping. Most species of primate have thumbs, although there are a small number of arboreal species that have only four fingers on each hand, such as the mantled guereza colobus monkey (*Colobus guereza*). In these cases, the hook-shaped hand is used effectively for rapid movement in the trees, and the leafy diet does not require complex manipulation of food items. For the majority that do have thumbs, a range of opposability exists. Strictly speaking, "perfect opposability" is defined as the act of placing the pad of the thumb fully and directly against the pad of a finger. This motion can be seen in many nonhuman primates, although the degree of placement may be less exact than what is demonstrated by humans. In general, Old World monkeys and apes exhibit a higher degree of thumb-finger opposability than the prosimians and New World monkeys.

All primates, except for bipedal humans, have a very mobile big toe that is separate from the other toes and can be used for grasping and holding (above left). Most primates, such as this chimpanzee (right), have a thumb, although the range of opposability found in each species varies. Old World monkeys and apes are capable of greater precision than the prosimians and New World monkeys.

The precision grip resulting from an opposable thumb has had an enormous adaptive impact on primate evolution, producing a range of behaviors that are unique among all other species. This level of dexterity allows a single grain of wheat to be separated from a pile of sand, a stray eyelash to be gently removed from an infant's face, and the manufacture and use of tools in ways that are simply too numerous to list.

Eyes

All species of primate have relatively large eyes and rely on vision as their primary sense. In general, olfactory abilities have been greatly reduced compared with visual skill. Primates always have forward-facing eyes situated so that the axes of vision overlap and are perceived by the brain as a single, three-dimensional image. This binocular style of perception facilitates accurate assessment of distance and depth, both of which are clearly relevant for animals that routinely move high above the ground and navigate in a complex environment. Except for the species that are active only at night, all primates see color.

All primates, like this gorilla, have forward-facing eyes that are situated to allow binocular, stereoscopic vision. All day-active species see color.

The importance of vision for primates is emphasized by the skeletal structures that function primarily to protect the eyes. The skulls of lemurs and lorises have a postorbital bar around the eye, while monkeys and apes, including humans, have a postorbital closure, which forms a bony cup around each eye.

Brain

The physical features that characterize primate brains are easy to identify, but they do not explain why or how primates exhibit a level of behavioral flexibility that is distinct from most other mammals. Neurobiologists are still searching for those answers, and some of the most current questions in the field will be addressed in Chapter 3, "Primate Intelligence."

Within the primate order, a large range of different body sizes exists. Of course, the hands, hearts, stomachs, and all other body parts vary as well on the basis of body size. It is simply expected that a 36-kilogram monkey will have a larger heart, for example, than a 4.5-kilogram monkey. The same logic applies to brains. The brains of larger primates are bigger than the brains of smaller primates. Since this is simply a measure of scale, the "absolute size" of the brain doesn't mean very much. However, a simple calculation based on the weight of the brain divided by the weight of the body gives a more interesting measure, referred to as relative brain size. The relative size of primate brains tends to be larger and their complexity tends to be greater than expected when compared with other similarly sized mammals, although this generalization cannot be applied reliably throughout the order. An important distinction, however, is that all primates consistently have large relative brain sizes during the fetal stage.

Primate brains can be best characterized by focusing on trends in structural organization. All primate brains follow a similar pattern of increasing complexity, and a larger size is favored for the parts of the brain that have evolved relatively more recently, such as the neocortex. The neuroanatomy of the brain clearly demonstrates that vision is the predominant sense and has been strongly emphasized throughout the evolution of the order. Olfaction and taste have correspondingly decreased in their neuroanatomical representation. All primate brains have mechanisms supporting sophisticated motor skills and sensitivity for the hands and feet and a prehensile tail for those species in which a tail is present. Although differences in neural complexity are evident among the primates, it is overwhelmingly clear that an evolutionary tendency toward sophisticated and complicated brains is a distinction for the entire order.

Life Stages
Primates share a number of distinctions related to their offspring. Every individual goes through a series of life stages following the pattern of infant, juvenile, adolescent, and then adult. Sexual maturity emerges during adolescence, prior to full physical development. Among primates, adolescence occurs later in life than among other animals. For animals of their size, primates have a long period of gestation, and most species (but not all) commonly give birth to singletons. Offspring have a very long period of dependence, with a corresponding high parental investment. In most cases, offspring rely exclusively on the mother for all direct parental care. However, there are species where this level of direct care is shared by both parents or by the extended family group, as in golden lion tamarins (*Leontopithecus rosalia*). Some species, such as chimpanzees (*Pan troglodytes*) and humans (*Homo sapiens*), maintain strong relationships between offspring and parent that may last for a lifetime. All primates have long lives for animals of their size.

ARE MONKEYS AND APES THE SAME THING?
The general term *primate* includes all members of the order. Therefore, every species of prosimian, monkey, and ape is correctly referred to as a primate. The terms *monkey* and *ape* are used more specifically. However, common names, which may be descriptive but are not always accurate, can create confusion. For example,

All primates have a long period of dependence compared with most other animals, and mothers are the first and most important social relationship for every primate. In many species, such as olive baboons (left) and mountain gorillas (right), female offspring may maintain a relationship with their mother for a lifetime.

Barbary macaques (*Macaca sylvanus*) are monkeys, but they are frequently called Barbary apes. A much smaller number of species are included in the category known as apes. These are divided into two groups, the "lesser apes" and the "great apes." The lesser apes include only the members of the family Hylobatidae, the arboreal gibbons of the Far East. The great apes include only the family Hominidae, which comprises orangutans (*Pongo* spp.), gorillas (*Gorilla* spp.), chimpanzees (*Pan troglodytes*), bonobos (*P. paniscus*), and humans (*Homo sapiens*). Primate groups such as the macaques, tamarins, and baboons are all correctly referred to as monkeys, not apes. Chimpanzees, gorillas, and gibbons are examples of apes and should never be called monkeys.

A number of anatomical features distinguish the apes from most monkeys. However, these are valuable only as generalizations, since there are exceptions for each. The great apes are very large compared with monkeys, although the lesser apes are not. This is the most likely origin for the distinction of "great ape" and "lesser ape." None of the apes have tails. Most monkeys do have obvious tails, but a small number of their species have extremely small tails or no tail at all. The Barbary macaques, mentioned above, are a good example of this, which helps to explain their confusing common name. All of the apes (except humans) have arms that are

There are many species of monkey that have very obvious tails, such as the mantled colobus (left), while others do not. Orangutans (right), like all of the apes, never have tails.

longer than their legs, and most monkeys have arms and legs that are of roughly equivalent length. In a typical quadrupedal posture, the shoulders and hips of monkeys are about the same distance from the ground. In the same pose, the shoulders of gorillas, chimpanzees, bonobos, and orangutans are higher than their hips, forming a sloping posture. Humans and gibbons rarely stand quadrupedally; instead they usually assume a bipedal upright position with the shoulders directly above the hips. All of the apes have a wider range of motion in their shoulder joints compared with monkeys. When in a bipedal standing position, apes are capable of placing their arms in positions that the majority of monkeys cannot. The New World spider monkeys (*Ateles* spp.) are the exception to this rule. They are the ecological counterparts of the Asian lesser apes, with very similar shoulder anatomy and range of motion. These few generalizations help to distinguish between monkeys and apes, but each has obvious exceptions.

Fortunately, cognitive testing of monkeys and apes has revealed one reliable distinction. In experiments designed to test mirror self-recognition, all of the apes have demonstrated the ability to recognize their reflections. Despite years of study, no monkeys have shown this same capacity. Future research into the general mental abilities of monkeys and apes is likely to reveal other distinctions as well.

Although much research is still to be conducted, there appears to be a significant difference between the mental abilities of monkeys and apes. While some general anatomical features may be useful for distinguishing between these two groups of pri-

Red-bellied lemurs show the typical posture of vertical clingers and leapers.

mates, emerging evidence suggests that the structure and related functions of the brain may be more valuable for drawing conclusions about overall degree of similarity.

HOW DO PRIMATES MOVE?

Primates move in distinct ways that are adapted to specific environments and natural histories. Interaction between individuals and their environment is constantly influenced by natural selection. Over time, behavioral and anatomical characteristics that promote maximum success persist in populations. The different forms of locomotion that are demonstrated by separate groups of primates provide excellent examples to illustrate this point.

Primates' styles can be lumped into four general categories. Vertical clingers and leapers spend much of their time in the trees and keep the trunk of their body in an upright position as they move or rest. In the trees, they propel themselves forward or backward by pushing off with their legs, which are longer and stronger than their arms. They land feet first and then grasp with their hands. When on the ground, these primates move in a rabbitlike fashion. If they feel threatened, they can rise to

Sifakas use their long, powerful legs, which are easily seen here, to make impressive leaps between trees. When traveling on the ground, they may stand upright and propel themselves with a hopping, bipedal style of movement.

an upright posture and bounce along rapidly using only their legs. Many of the prosimians fall into this category of movement, and sifakas (*Propithecus* spp.) are an especially good example.

The overwhelming majority of primates are quadrupedal, meaning they use their arms and legs simultaneously as they locomote, although this form of movement is expressed very differently from species to species. For most of this group, the length of the legs is very similar to the length of the arms. The lorises of the *Loris* genus, for example, have hands and feet that are adapted for gripping like a vise but are incapable of fine manipulations. They move slowly and deliberately, securely clamping onto branches as they travel. By contrast, macaques (*Macaca* spp.) have hands that are capable of impressive precision and a wide range of movement. They are at ease exploiting almost any habitat, whether in the trees or on the ground, and clearly do not fit into the "slow and deliberate" category. They are capable of using the quadrupedal style of movement in almost any situation and at a range of speeds. Unlike the lorises that always grip tightly as they move, macaques extend their fingers and toes while walking along branches or running on the ground. In this digitigrade posture, their body weight is distributed along the front half of the hands and feet. When macaques move vertically, they use their arms first, followed by their legs. Both the hands and the feet grasp strongly during any climbing movements. Leaping quadrupeds, such as the arboreal colobus monkeys (*Colobus* spp.),

Quadrupedal movement, as shown by the posture of this olive baboon (left) and by the stride of a mountain gorilla (below), is the most common form of locomotion found among primates.

move very differently from the vertical clingers and leapers previously described. Colobus frequently jump across gaps in the forest canopy as they move from tree to tree. When they land, they stabilize themselves using their arms first or using the arms and legs simultaneously.

Although quadrupedalism is common as a style of movement within the order, the slightly different expressions of it highlight the ways in which different species have evolved to fill specific niches in their environment. The dynamic interaction between locomotion, behavior, anatomy, and environment has resulted in several variations on a theme that are all successful and have become predominant within the order.

Proboscis monkeys typify leaping quadrupeds. This individual makes a leap between trees during travel through the forest canopy.

The last two forms of movement are relatively rare and involve clear anatomic specializations. Brachiation is the form of locomotion in which a primate is suspended only by its arms, without the legs supporting any weight at all. The movement of brachiation follows a symmetrical pattern in which one hand grasps and holds as the body swings beneath it like a pendulum. As the first hand releases, the other reaches to grasp, and the body continues to be propelled forward. During rapid brachiation, forward movement continues between each handhold, and the animal continues to travel in midair without touching anything. Primates that are able to brachiate all have very long arms, exceptionally mobile shoulder joints, long fingers, and a reduced thumb. As they move, the hand is used like a hook, with only the four fingers coming into contact with branches. While a thumb is present, its small size and distance from the other fingers limits how it may be utilized. While perfectly adapted for brachiation, it is not very useful for manipulating objects.

The unrivaled masters of brachiation are the gibbons (*Hylobates* spp.), native to the tropical forests of the Far East. These monogamous apes, which range from about 6 to 15 kilograms depending on the species, rarely descend to the ground and subsist on a diet that contains a very large percentage of fruit. In addition to the typical anatomical features mentioned for brachiators, gibbons also have a short trunk and lumbar area, which reduces the strain that may occur from repetitive bending while hanging and reaching simultaneously. The athleticism and grace exhibited by gibbons during brachiation is simply breathtaking. With no apparent effort, they flawlessly execute gymnastic movements that would require years of training for skilled human athletes. The speed and precision of their movements

The body proportions of this gibbon are typical for all of the species within the genus *Hylobates*.

during brachiation are difficult to comprehend but can be illustrated by reports of gibbons pursuing and catching birds as they fly through the treetops.

In addition to these "true" brachiators, a small number of other species are considered "semi-" or "modified" brachiators. Among the platyrrhines, spider monkeys (*Ateles* spp.) combine the movement of their arms and prehensile tail in a suspensory style of locomotion that is a variant of true brachiation. All of the great apes are capable of modified brachiation as well. However, it is never a dominant form of movement and is more frequently demonstrated by juveniles. Among the great apes, orangutans are the most accomplished semibrachiators. In the trees, their predominant style of travel is described as quadrumanous, referring to the way in which the hands and feet are used interchangeably. Along unobstructed horizontal pathways, orangutans perform modified brachiation with ease. However, their bulk prevents them from performing anything that resembles acrobatics, and they maintain a firm grasp at all times. Although fluid and graceful, their movements conjure an image of strength and raw power. By contrast, gibbons appear almost delicate as they maneuver, as though barely touching their surroundings.

Masters of true brachiation, gibbons literally fly during travel through the treetops. Note the upright body posture that is maintained compared with the forward tilt of the leaping quadrupeds.

Bipedalism is the final form of movement as well as the rarest. It is habitually practiced by only one species, *Homo sapiens*. While many other primates are able to stand on their legs and walk or run for short distances, the stride exhibited by humans is unique within the order. Continuous erect posture and alternating balance on each foot require a variety of anatomic specializations. These are primarily located from the waist down, which frees the arms for other tasks during locomotion. Humans have legs that are longer and significantly stronger than the arms. They also have a spine with an S-shaped curve, which moves the trunk forward and places the mass of the body over the hips. The position and shape of the pelvis promote balance and lower the body's center of gravity. The attachment points of the muscles at the hip also assist with balance and facilitate placing the legs behind the trunk of the body. The largest, strongest, and heaviest bone in the human body is the femur, which has significant adaptations related to bipedalism. The head of the femur is very big and carries the full weight of the body during normal movement. Viewed straight on, it is apparent that the femurs angle slightly inward toward the knees. During walking, this results in a smooth stride, since each leg is situated al-

Orangutans have tremendous flexibility and live in the same habitat as the gibbons. While orangutans may travel using only their arms, their body weight obliges them always to keep one hand or foot attached as an anchor. As a result, they are considered "semibrachiators."

most directly below the middle of the body as the foot touches the ground. The other primates lack this specialization and tend to rock from side to side as they move bipedally. The human foot is also distinct. It is poor at grasping and fairly rigid as well. With its short toes and flat appearance, it functions best at providing propulsion during walking and running.

Bipedalism presents an interesting evolutionary mystery. Compared with quadrupedalism, it is slower, less agile, and may require more energy in some instances. Therefore, it must have conferred some pronounced benefit in order to persist in the population of our early ancestors. Possible advantages may have been related to tool use, efficient foraging, social bonding, or communication. Of course, benefits may have accumulated from all of these areas of behavior. Although the exact circumstance that initially promoted bipedalism may never be identified, it is certain that this style of movement has contributed heavily to our success and influence as a species.

HOW CLOSELY ARE HUMANS RELATED TO OTHER PRIMATES?

Taxonomic relationships are based on how closely or distantly species are related. Similarities at the very broadest levels of classification are usually intuitive, such as plants versus animals, vertebrates versus invertebrates, or birds versus mammals. However, the exact relationship among the species within an order may not be as obvious. Since relationships within an order are closer than those between orders, humans have a higher degree of relatedness to all other species of primate compared with any other type of organism. Within the order Primate, humans and the prosimians have the greatest evolutionary distance from each other and therefore the lowest level of relatedness. In increasing levels of relatedness to humans, the New World monkeys are next , followed by the Old World monkeys, and then the gibbons. The closest relation exists between humans and the other members of the family Hominidae, the great apes.

In the past, relationships among species have primarily been determined by examining similarities in appearance. While anatomical comparison always has value, more recent techniques that allow for investigation at the molecular level have become increasingly important. DNA, the biological blueprint for each individual, has been particularly useful. DNA comparison is one of the most commonly used and reliable methods for understanding the biological relationship and genetic closeness of any two species. The level of homogeneity is generally expressed with a percentage that represents overall genetic similarity as well as evolutionary distance.

Comparing the DNA of the five species within Hominidae provides clear evidence to illustrate the level of relatedness between humans and the other great apes. Not surprisingly, the lowest level of variation in DNA is between chimpanzees (*Pan troglodytes*) and bonobos (*P. paniscus*), members of the same genus. The DNA of these two species is 99.3 percent the same. The next closest relationship exists between modern humans (*Homo sapiens*) and the members of *Pan*. Humans, chimpanzees, and bonobos share 98.4 percent of the same DNA sequence. Gorillas (*Gorilla* spp.) share 97.7 percent of their DNA with *Pan* and *Homo*. Orangutans, the Asian great apes, share 96.4 percent of their DNA with *Pan*, *Homo*, and *Gorilla*.

This information clearly reveals that the closest evolutionary distance exists

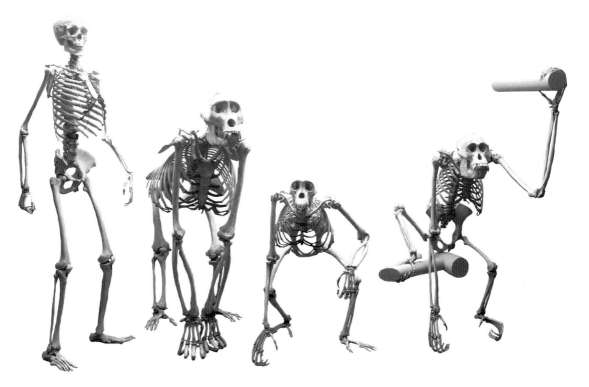

These skeletons of the members of the family Hominidae show the overall similarity in body plan that exists among these species. However, differences related to locomotion are clearly evident. Humans (far left) are the only species that is primarily bipedal, while orangutans (far right) are the only species that is primarily arboreal, depending heavily on the arms for locomotion. Gorillas (center left), chimpanzees (center right), and bonobos (not pictured) are terrestrial quadrupeds. (Photograph by Chip Clark, National Museum of Natural History, Smithsonian Institution)

Humans and chimpanzees share remarkable physical similarities and are more closely related than gorillas and chimpanzees. Decades of research by Jane Goodall, pictured here with a wild chimpanzee, have revealed previously unknown behavioral similarities as well.

among humans, chimpanzees, and bonobos. Therefore humans, not gorillas, are the nearest relatives to the chimpanzees and bonobos. However, the 98.4 percent of DNA that humans share with chimpanzees and bonobos does not imply a corresponding level of overall equivalence. Rather, it points to the profound consequences that a mere 1.6 percent difference in DNA can create. While humans and the other great apes have notable physical, behavioral, and cognitive similarities, it is *Homo sapiens* that has evolved to be the dominant species on the planet. This position has profound benefit, and our global influence cannot be denied. However, it also carries a massive responsibility. Human behavior will ultimately determine the fate of our closest evolutionary relatives.

DID HUMANS EVOLVE FROM APES?

Of all the questions that pertain to primates, the one having to do with evolution undoubtedly generates the strongest emotional reaction. In fact, the arguments that began in 1859 with Charles Darwin's publication of *On the Origin of Species* are still being debated today. The roots of the controversy are easy to understand, since the idea that species have evolved over time directly challenges the persistent notion that humans were created in their modern form. Common misunderstandings about the process of evolution only add fuel to the fire.

In Darwin's time, the process of evolution was based on an understanding of three fundamental principles. The first was that all species in a population have slightly different characteristics. The second principle stated that offspring inherit some characteristics from their parents. The third acknowledged that as a result of these varying characteristics, some individuals are more successful than others and have a greater number of offspring. These fundamentals focus on variation, heredity, and differential reproduction, which is also called fitness. In Darwinian evolutionary theory, fitness is not a measure of physical health but rather refers to the number of surviving offspring left by an individual. Natural selection is the mechanism that determines which characteristics are best suited to the environment and will persist in succeeding generations. These variations, which may be either physical or behavioral, will be favored unless environmental conditions change. Natural selection is also commonly described by the phrase "survival of the fittest." While the transmission of traits from one generation to the next was recognized in Darwin's era, genes had yet to be identified as the biological vehicle that carried information from parents to offspring. Today, the science of genetics has enormous influence on our understanding of the evolutionary process.

The fundamentals of Darwinian evolution have been refined to include the important concept of genetic variation. Alternate forms of the same gene are called alleles, which are the basis for the range of characteristics that may appear in any individual. Therefore, heredity involves the transmission of alleles from parent to offspring. Differential reproduction, under the influence of natural selection, promotes the persistence of certain alleles that ultimately increase the fitness of their owners. Evolution is most simply defined as a change in the frequency of alleles in a population over time. The process of evolution is the unifying principle that is the basis for all of biology and can be applied to any living organism.

Evolution is not goal directed. Rather, it is a biological mechanism that results in change within populations. Over time, segments of a population may become increasingly isolated from each other, physically, behaviorally, and/or geographically. If these divisions are dramatic or are in place for long periods, individuals with the same ancestors may become quite different from each other. These populations may be sufficiently distinct to be considered different species. The process of speciation may be slow or relatively rapid, depending on a variety of features including environmental pressures, rate of reproduction, and rate of genetic mutation. Additionally, species do not necessarily emerge in an orderly or linear fashion. The traditional concept of evolution as a "tree," with one species neatly branching into another, is misguided. A more accurate metaphor would be an evolutionary "bush," with many, many branches that begin and end in a confusing tangle. The fossil record confirms that human evolutionary history has proceeded in this way.

The common assumption that humans evolved from one of the species of modern great ape is completely wrong. All of the species in Hominidae do share a common ancestry, but each has a totally distinct evolutionary pathway. Approximately 25 million years ago, the ancestors of the Old World monkeys diverged from the ancestors of Hominidae, and the ancestors of modern gibbons split from Hominidae approximately 18 million years ago. The orangutans of Asia began a distinct lineage within Hominidae approximately 14 million years ago. The remaining species formed the ancestral core for all of the African great apes, including humans. The gorilla line formed its own branch approximately 7 million years ago. Humans were the next lineage to emerge, approximately 6 million years ago, and the chimpanzee-bonobo split occurred around 3 million years ago. Despite common misperceptions, the human lineage was not the conclusion to great ape evolution but was simply another of the multiple branches.

It is also important to remember that the modern species in Hominidae are only the current success stories. The human lineage has included many other species, and *Homo sapiens* is the only member of our direct lineage that has persisted to the present day. The desire to find the "missing link" that represents the single intermediate form between humans and the other great apes ignores the probability that such an animal never existed. Undoubtedly, many intermediate species existed throughout the evolutionary history that has involved humans and the other great apes. The range of characteristics shown by these fossil remnants does not reveal a steady march toward *Homo sapiens* but provides extraordinary evidence that attests to the forms and adaptations that have been promoted by the process of evolution.

WILL CHIMPANZEES EVOLVE INTO HUMANS?

Charles Darwin was not the first person to propose a scientific explanation for the diversity of life on Earth. Around 1800, Jean Baptiste Lamarck offered his theory, which, like Darwin's, focused on the variety of characteristics that were present in a population of organisms. Lamarck differed by suggesting that these features were acquired in response to a specific environmental demand. The long neck of the giraffe is the classic example. A Lamarckian explanation would begin with the idea that the ancestors of giraffes had short necks. These individuals constantly stretched their necks reaching for foliage high in the trees and, as a result, had offspring with longer necks. The same phenomenon would occur in succeeding generations, resulting in the long-necked giraffes that we know today. Lamarck believed that the environment presented a challenge and that the solution created a need for a particular characteristic (such as a longer neck). An organism's efforts affected its physical structure, which would then be inherited by offspring.

Darwinian evolution clearly rejects those arguments. It is now understood that the process of evolution relies on random changes rather than a response to a specific need or desire. New characteristics that are beneficial will persist in a population, and the rest will be eliminated or reduced. The idea of chimpanzees evolving into humans is a clear case of Lamarckian logic. In this situation, the characteristics of *Homo sapiens* would be the desired result for chimpanzees, which is impossible, since the process of evolution is not goal directed.

The question also assumes that humans are an endpoint of evolution and that other species are evolving into a finished product as well. The evolution of any species is continuous, having no defined ending except for extinction. Humans are by no means a finished product and are not the "most evolved." All living species are moving in their own evolutionary direction. All are successes, although some species will exist longer than others.

The remarkable similarities between humans and chimpanzees (as well as bonobos, gorillas, and orangutans) are primarily the result of a common ancestry. Additionally, the physical, mental, and behavioral features that are shared are not "human" features that the great apes have borrowed. Rather, they are characteristics that have been successful for each species and have persisted in their respective

"If we look straight and deep into a chimpanzee's eyes, an intelligent, self-assured personality looks back at us. If they are animals, what must we be?" —Frans B. M. de Waal (1982, 18)

populations. While it is certainly true that chimpanzees look like humans, it is also true that humans look like chimpanzees.

WHAT ARE THE SMALLEST AND LARGEST PRIMATES?

The appearance and behavior of every species is under the constant influence of natural selection, and size can be a very important variable for determining how successfully a species survives and reproduces. Body size can also be a good predictor for certain trends in behavior and natural history. The smallest and largest species of primate fit into this model well.

The smallest primate is the pygmy mouse-lemur (*Microcebus myoxinus*), which, like all lemurs, is found only on the island of Madagascar off the east coast of Africa. When the pygmy mouse-lemur is fully grown, the combined length of its head and body of averages 61 millimeters in length, and the tail is over twice as

The mouse-lemur, about to pounce on a cricket here, is the smallest of all primates.

long as the body, measuring 136.2 millimeters. The weight of this wee primate averages 30.6 grams, about the same as a big strawberry. Females may be slightly larger than males. Mouse-lemurs live in forests and subsist on a diet that includes fruits, leaves, flowers, nectar, tree sap, insects, spiders, and sometimes frogs and lizards. Like most tiny mammals, mouse-lemurs rely on behavioral strategies for protection from predators and competitors. In order to be as inconspicuous as possible, they usually travel alone or in very small groups and spend most of their time in the trees. They are nocturnal, foraging for food under the cover of darkness. Their secretive lifestyle makes them particularly difficult to study in the wild.

The largest primate is the gorilla (*Gorilla* spp.). In the wild, gorillas occur in a limited range that is concentrated across the middle section of Africa, beginning in Nigeria and ending in Rwanda, Uganda, and the Democratic Republic of the Congo (formerly Zaire). In the wild, adult male gorillas may have a combined head and body length of 1,700 millimeters and weigh around 160 kilograms. Gorillas exhibit extreme sexual dimorphism, with adult males growing to twice the size of adult females. Given their very large size and strength, gorillas have no natural predators (except for humans armed with weapons) and can afford to be conspicuous in their environment. As a result, social groups may contain more than twenty individuals, who spend large amounts of the day foraging for vegetation and fruits. Wild gorillas have never been observed to hunt or eat meat. While large size may confer benefits, it also creates risks. Gorillas require large amounts of forest in which to forage for food, and that frequently places them in competition with local human populations. Currently, all populations of wild gorillas are in serious decline and at great risk of extinction.

The gorilla is the largest of all primates. Extreme sexual dimorphism is the norm for this species of great ape; adult males may be twice the size of adult females.

WHAT DO PRIMATES EAT?

Almost all species of primate consume some form of vegetation or fruit in their normal diet. Most also eat flowers, seeds, bark, nectar, and other plant parts. A smaller number regularly prey on invertebrates such as larvae, ants, and crabs, and others eat small vertebrates like frogs and fish. Tarsiers (*Tarsius* spp.) are a unique case and subsist entirely on invertebrates and small vertebrates, such as snakes and lizards. They are the only primate that can be accurately labeled a faunivore. A very small number of species regularly hunt and eat mammals (see *Do Any Primates Hunt for Meat?*). Of all types of animals, primates have the most varied and complex diets. No matter what the item, if it can be chewed and swallowed, some type of primate is bound to eat it.

Some species may be limited in what they consume, but this is usually the result of some anatomical adaptation or specialization. For example, the aye-aye (*Daubentonia madagascariensis*) has a diet that is heavily dependent on insect larvae. It has a highly specialized, bony middle finger with a claw on the tip that it uses to locate its prey inside tree trunks by tapping and listening. Aye-ayes use their rodentlike incisors to gnaw into the wood and then extract their prey with their skeletal finger. This finger is also inserted into coconuts to withdraw the contents. The New World tamarins have very agile and sensitive fingers that are used for a specialized type of foraging called micromanipulation, in which they blindly probe tree crevices and grab anything that might be edible. While potentially risky, this behavior allows them to prey on items that might literally be overlooked.

The sacculated stomachs of the Colobinae are simply unable to handle the wide range of foods that other primates easily consume. However, this restriction is

The crest on an adult male gorilla's head serves as an attachment point for the muscles and tendons of the jaw. The strength of these muscles allows males to process the large amounts of vegetation they need to meet their daily nutritional needs.

actually a specialization that allows them to digest and extract nutrients from leaves that are useless for most other species of primate. Although their variety of foods is somewhat limited, abundance is always assured. The only noncolobine that has a similar lifestyle is the gelada (*Theropithecus gelada*), which subsists almost entirely on grasses. The superavailability of this resource allows geladas to have a very flexible social system with limited competition over food. The prominent sagittal crest on the head of adult male gorillas and orangutans is associated with food consumption. Because males are much larger than females, they process relatively greater amounts of food on a daily basis. The mandibular muscles that are used for chewing are attached high on the crest, providing power and endurance (which also comes in handy during male-male aggressive encounters).

It is frequently assumed that the morphology of primate teeth provides an easy key for identifying dietary preferences. Although some clear adaptations do exist (such as the gnawing teeth of the aye-aye), the relationship between teeth and diet is not always direct. While it may be true that certain species mainly eat fruit, the texture, skin, and composition of all the fruits that are eaten may vary widely from one species to another. The same can be said for a diet that is heavy with invertebrates. Therefore, assumptions about teeth and diet must be made cautiously and with copious information about food preferences from long-term field studies. The most accurate statement about diet and teeth is that primate dentition is generalized, and the feeding habits of most primates are very adaptable to changing circumstances.

DO ALL PRIMATES HAVE TAILS?

Most primates sport a very obvious tail. Those that don't sport an unmistakable tail have either a very small tail or the vestiges of one. For those species that do have conspicuous tails, the size and the ways that it can be utilized may vary greatly. "Tails" can be grouped into four distinct categories on the basis of their general appearance and how they function.

The largest category includes tails that are very conspicuous but have little controlled movement associated with them. The species with this type of tail use it primarily for balance and propulsion during locomotion, especially in jumping and leaping. The colobus monkeys (*Colobus* spp.) fit nicely into this group, and the mantled guereza (*Colobus guereza*) provides an excellent example. These African foliovores live almost exclusively in the treetops of the rain forest and make spectacular leaps when traveling between gaps in the canopy. Adults measure approximately 120 centimeters from head to tip of the tail, with the body making up only half of that length. Their long tails have a luxuriant mass of hair and provide stability during leaps between trees. However, colobus monkeys have very little direct control over the movement of their tails. In areas where colobus monkeys are prey for chimpanzees (*Pan troglodytes*), they have been seen to protect themselves by holding their tail in their hands, away from the reach of the apes hunting on branches below. Other primate species may exert a bit more control over the movement of their tails. Lion-tailed macaques (*Macaca silenus*) have bodies that are around 50 centimeters long, with slender tails of about 25 centimeters in length. They have the ability to alter the position of their tail to communicate information about social status or intention. For example, a tail that is held in an upright arc over the back indicates dominance and confident social standing. By contrast, a low-ranking lion-tailed macaque would likely hold his tail in a limp position that touches the ground. The tails of infants and juveniles also serve a useful function, since mothers frequently keep track of their offspring by holding on to their tails like a leash.

The second descriptive category for tails includes only those that are prehensile,

Many species of monkey with conspicuous tails have very little control over them, such as these olive baboons.

meaning that they are able to hold, grasp, and function as a fifth extremity. This amazing adaptation evolved only in the New World monkeys and occurs in the woolly spider monkeys (*Brachyteles* spp.), spider monkeys (*Ateles* spp.), woolly monkeys (*Lagothrix* spp. and *Oreonax flavicauda*), howler monkeys (*Alouatta* spp.), and the capuchin monkeys (*Cebus* spp.). Prehensile tails are sensitive, muscular appendages that are useful for locomotion, in foraging, and in expressing a variety of social behaviors. All prehensile tails (except those of capuchins) have a naturally hairless area on the ventral aspect near the tip, called a friction pad. This area is similar to the palm of the hand or sole of the foot and has a distinctive print. These prints can be used to distinguish different genera, species, and even individuals. Prehensile-tailed monkeys can comfortably use their hands to forage for food or interact with other individuals while hanging upside down and secured only by the grasp of the friction pad. While traveling, the tail alternates with the hands to hold on to branches or can be wrapped around a trunk in order to descend from a tree safely. Adult female spider monkeys frequently use their tail to form bridges for in-

The prehensile tails of spider monkeys function as another hand or foot and have a wide range of uses. This individual has a secure hold with its tail, effectively freeing the hands for foraging.

Capuchin monkeys are unique, since they are the only species outside the family Atelidae that has a prehensile tail. Unlike the atelids, capuchin tails are fully covered with hair and are not used for locomotion. Their main function is to provide support during foraging.

White-handed gibbons, like all of the apes, lack any external tail.

fants and juveniles. Upon encountering a gap in the forest, adults will grasp a branch with their tail and then leap to the next tree, securing themselves with their hands. They maintain a grip with both the tail and the hands, forming a bridge with their body. Once their offspring have safely moved across the gap, the adult simply releases the branch held by the tail and continues on her route.

The third category of tails includes those that range from very short to nothing more than barely visible stumps. The species with these tails appear to be randomly distributed throughout the order, with no particular features in common except for their diminutive appendage. The prosimians in this category are the angwantibos (*Arctocebus* spp.), the potto (*Perodicticus potto*), the false potto (*Pseudopotto martini*), all members of the genera *Loris* and *Nycticebus*, and the indri (*Indri indri*). The lone examples from the New World monkeys are the uakaris (*Cacajao* spp.). The Old World monkeys included in this group are the drills and mandrills (*Mandrillus leucophaeus* and *M. sphinx*), the aptly named pig-tailed langur (*Simias concolor*), and many species of macaque (*Macaca* spp.). It is important to mention that within these species, there can be some variation among individuals. Unlike the previous

Mouse-lemur, *Microcebus* spp., Madagascar

De Brazza's monkey, *Cercopithecus neglectus*, Africa

Lion-tailed macaque, *Macaca silenus*, India

Proboscis monkey, *Nasalis larvatus*, Borneo

Siamang, *Hylobates syndactylus*, Sumatra,
Malay Peninsula

Ring-tailed lemur, *Lemur catta*, Madagascar

Olive baboon, *Papio anubis*, Africa

White-headed capuchin, *Cebus capucinus*, Central and South America

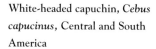

Mountain gorilla, subspecies *Gorilla beringei beringei*, Rwanda, Uganda, Democratic Republic of the Congo (formerly Zaire)

Orangutan, *Pongo* spp., Sumatra and Borneo

Chimpanzee, *Pan troglodytes*, Africa

two categories, none of these very short or stumplike tails are utilized for any specific function.

The last category includes only those species that have absolutely no external tail. The only members of this group are the lesser and great apes, including humans. However, a tail is apparent during fetal development among these species. By the end of the first six weeks of gestation, humans have a tail that contains 10 to 12 vertebrae and cannot be distinguished from the appendage found in species that retain their tails throughout gestation. By the end of eight weeks of development, the human tail regresses and the vertebrae reduce and fuse to form the coccyx. The coccyx, also commonly called the tailbone, is the vestigial skeletal remnant of a tail and is found in lesser apes and great apes, including humans. The coccyx is a series of small bones that form a point, which is situated on the back of the pelvis. On rare occasions, the normal process of regression does not occur, and human infants have been born with external tails. The coccyx is one remnant of the evolutionary heritage shared by all primates.

HOW LONG DO PRIMATES LIVE?

Primates that live in the wild endure constant challenges to their survival. Basic environmental conditions such as temperature and availability of food and water affect life spans. Disease, parasites, and predators also stress individuals and local populations, and deaths that occur in primate groups follow a predictable trend. Mortality is relatively high for infants, low for juveniles, and then increases for adults as they age. While this is true for the populations as a whole, the situation for males versus females is dramatically different. In many species, male primates are expected to have much shorter life spans than females. During adolescence and through adulthood, a much larger number of males die compared with females of the same age. One factor that can be used to predict this statistic is the frequency rate of males' migration among groups. In many primate social systems, young male primates are rejected by their natal group during adolescence and are forced to spend time as solitary bachelors, eventually attempting to join a group with unrelated females. During this transition, males lose many of the benefits associated with living socially and face much higher risks of predation. Assimilation into a new group can be very dangerous, and young males may be attacked and fatally injured by resident males. If immigration is successful, challenges from rival males are guaranteed throughout adulthood. The reality of this bleak scenario favors males with the best behavioral and physical qualities and promotes the survival of their genes into the next generation.

Broader trends associated with longevity are also evident. As is the case for most mammals, a correlation exists between size and life span. Although exceptions exist, smaller primates usually have shorter lives than larger ones. In the wild, very small primates such as the tarsier (*Tarsius* spp.) are estimated to live around 10 years, macaques (*Macaca* spp.) may live around 25 years, and great apes have been documented to live into their late 30s. In order to collect accurate information about longevity in the wild, we must reliably identify individuals and follow them from birth to death. This may be extremely difficult as emigrations occur, social groups change, or habitats are disrupted. Additionally, the life span of some primates, such as gorillas, may simply be longer than the length of a continuous field study.

Information from captivity is very helpful for answering questions about the maximum age that primates may reach. However, in these settings, it is important to recognize that many of the normal factors that shorten lives in the wild have been eliminated. Primates that live in zoos, sanctuaries, or other managed settings are routinely protected from hunger, extremes in temperature, disease, aggression, and predation. Not surprisingly, captive primates almost always live longer than their wild counterparts. For example, macaque ages may reach into the late 30s or early 40s, chimpanzees have been known to live into their 50s and early 60s, and humans (*Homo sapiens*) also enjoy greatly extended life spans when protected from these same risks.

WHICH PRIMATES ARE THE FASTEST?

The physical adaptations that promote an arboreal lifestyle are not very compatible with moving quickly on the ground. Therefore, most primates are not particularly fast compared with four-legged terrestrial animals. However, the gibbons (*Hylobates* spp.) are the clear masters of speed when moving through the forest canopy. As the true brachiators, gibbons use their arms to move with remarkable agility, athleticism, and precision. However, uninterrupted pathways high in the trees are uncommon, so it is more difficult to assess the true speed of brachiation compared with speeds of running on the ground.

Patas monkeys (*Erythrocebus patas*) are absolutely the fastest of the terrestrial primates. These monkeys live in open grasslands throughout the central portion of Africa and rarely spend time in trees, except for sleeping. Adult males weigh between 4.0 and 7.0 kilograms, and females may be half that size. Their main defense against predators is speed. Patas monkeys move quadrupedally, using the fingers (not the palms) of their hands and the soles of their feet. They have been documented to run as fast as 55 kilometers per hour on a flat stretch of open savanna. This ability rivals the racing speed of greyhounds and thoroughbred horses. The fastest human sprinter moves at approximately 32 kilometers per hour.

The patas monkey is the fastest primate on the ground, running at speeds that rival those of greyhounds and racehorses.

HOW STRONG IS A GORILLA?

Wildly exaggerated depictions of the gorilla have always focused on their incredible strength and savage behavior. Historical references about gorillas may have occurred as early as 500 B.C.; however, consistent reports did not begin to reach the Western world until the mid-1800s. Early explorers returned from Africa with fantastic tales about gorillas. One of the earliest of these, from Paul Belloni du Chaillu, is characteristic of the way in which gorillas were portrayed. Upon encountering an adult male gorilla in the forest, du Chaillu writes: "He stood about a dozen yards from us, and was a sight I think I shall never forget. Nearly six feet high (he proved four inches shorter), with immense body, huge chest, and great muscular arms, with fiercely-glaring, large, deep grey eyes, and a hellish expression of face, which seemed to me some nightmare vision; there stood before us the king of the African forest." Du Chaillu continued, "He reminded me of nothing but some hellish dream-creature—a being of that hideous order, half man, half beast, which we find pictured by old artists in some representations of the infernal regions" (quoted in Willoughby 1978, 39).

While certainly capturing the imagination of the public at the time, such stories as du Chaillu's unfortunately created a negative stereotype of gorillas that has persisted to this day. Many people still have the perception that gorillas are brutish and have the strength of 10 (or more) men. Studies from the wild and in captivity have documented that gorillas are generally shy and are more than likely to avoid confrontations whenever possible. Gorillas do have the ability to respond aggressively,

Humans marvel at the physical power displayed by gorillas. However, their strength has never been accurately measured, and the anatomical features that produce it are not well understood.

but this is usually limited to situations in which they are severely threatened or provoked.

The normal behavior of gorillas clearly supports the notion that they are much stronger than humans. During feeding, gorillas snap thick branches in half with no apparent effort and can split open stems of very heavy vegetation with only their hands or teeth. During dominance displays, adult males routinely pull small trees directly out of the ground and hurl them through the air. In captivity, gorillas have been known to remove metal plates, bend steel bars, and break welded joints as they investigate how an enclosure is constructed. Despite the reports from the wild and anecdotes from captivity, the limits of their physical strength have never been accurately measured.

While their superior strength is obvious, the exact mechanism for how it is produced is still unresolved. Body size is not an adequate explanation, since adult weights for gorillas can be comparable to those in the upper range for humans, but a huge disparity in strength still exists. It is more likely that a variety of factors contribute to produce the overall effect. Humans, a bipedal species, have about 50 percent of their muscle mass in the lower half of their bodies. The quadrupedal gorillas are the opposite, having about 50 percent in the upper half with comparatively little in their legs. These large areas of muscle are supported by a generous flow of arterial blood, which enhances performance. The attachment points for the muscle groups in the upper body may also promote better leverage. Although impossible to quantify, gorillas may also be less inhibited about using their strength. Humans living in urban settings, especially in developed countries, rarely have the need to develop or use the potential strength of their bodies. This psychological component may contribute little to the overall disparity, but it helps to explain our fascination and respect for the physical superiority demonstrated by gorillas.

.2.

PRIMATE SOCIAL BEHAVIOR

DO ALL PRIMATES LIVE IN FAMILIES?

Under normal circumstances, all primates spend at least a portion of their lives with one or more of their relatives. There is wide variation within the order, ranging from the relatively short time that prosimians spend with their young compared with the lifelong relationship that may occur between great apes and their offspring. The core social unit is some primate species may be only a mother and her offspring, while others live in congregations that may have hundreds of individuals.

In terms of understanding the social behaviors that correspond with these different types of groups, a common tendency is to use human norms as a reference point. This anthropocentric inclination is natural and provides one way to begin a discussion. After all, the human point of view is the only one that we can understand directly. Through the indirect means of observation and study, information about nonhuman primate social behavior can be gained. Much of what is observed may seem very familiar, and in fact humans share many (if not all) of the social behaviors exhibited by nonhuman primates. However, judgments about these behaviors must be made carefully, understanding that the human point of view does not necessarily provide an accurate explanation or interpretation of what may be occurring.

The social concept of the family is a good example. Traditionally, the core of the human family has been one adult male, one adult female, and their offspring. While this is the most common and familiar type of family for *Homo sapiens*, it is clearly not the only form that can be identified. Although less common, polygamous families have multiple adult females with one adult male. Polyandrous families are also known for humans, in which there are multiple adult males with one

Leaf monkeys (*Trachypithecus* sp.)

71

adult female. Other variations exist as well, and a host of factors affect the types of families that humans form. These factors may be cultural, religious, social, practical, emotional, and so on. At least some of them are unique to the human species and the formation of their families.

Use of the term *family* in regard to human society implies a range of elements or circumstances that should not necessarily be attributed to nonhuman primates. In order to promote accuracy and prevent unintentional anthropomorphism, it is much more useful to focus on specific topics related to social behavior than to try to force-fit all primate species into different categories of families. Investigation of these topics allows a much greater understanding of the forces and influences that govern the ways in which nonhuman primates function socially. This approach also permits a much more accurate comparison of human and nonhuman primate social behavior and assists in identifying both the similarities and differences that exist among species.

WHY DO PRIMATES LIVE IN DIFFERENT TYPES OF GROUPS?

A variety of forces, such as food and habitat availability and potential predators, influence the size, composition, and complexity of primate social groupings. As a result of ecological constraints, some species are best adapted for a solitary lifestyle, while others are successful living in groups that may have dozens or hundreds of members that travel and associate with each other. Variation exists within species as well, although to a much more limited degree. These factors are closely tied to the most important aspects of primate social life: food and sex (not necessarily listed in order of importance).

The most important challenges that all primates regularly face are centered on acquiring enough food and having the opportunity to reproduce. Of course, all types of food are not equal, and neither are all potential reproductive partners. Therefore, the challenge for each individual becomes focused on acquiring the most preferred types of food and the most desirable reproductive partners. Since both of these are usually limited in supply, conflict among individuals inevitably results. Social systems evolve in response to these demands and can be considered the best balance of "costs" and "benefits" for individuals. Any situation that involves living socially should have some direct benefit related to reproduction, protection, or access to resources.

The clearest advantage related to sexual behavior is simply easy access to other individuals. Primates living in groups don't have to search for sexual partners or wait to be found by another interested individual. Constant access also permits

Life in a large social group is a blend of costs and benefits. Social living may allow easy access to mates and effective protection from predators, but it also brings constant competition for all limited resources.

evaluation of potential partners, promoting selection of the most desirable mate. In a larger group, less desirable individuals have a better chance of finding another low-ranking mate, which, for a variety of purposes, may be better than no mate at all.

Social groups also provide greater security and protection for individuals and their offspring. Predators are more easily detected by larger groups because of the increased number of vigilant eyes and ears. A large group can cause confusion for a predator, reducing its ability to focus on a single individual. In any group, each member becomes less obvious and therefore reduces the risk of catching a predator's attention. If an attack does occur, many individuals can engage in group defense and possibly repel the predator. Threats and aggression may also come from rival groups. In these situations, group size is an essential component for defense and security.

The acquisition of food can also be facilitated by group living. In some cases, preferred foods may be obtained only through cooperative behaviors such as hunting. Foraging individuals may be able to spend more time feeding, since other group members are engaged in vigilance for predators. Groups are also effective at locating new food patches and can defend their resources against rivals.

Of course, not every aspect of social living is positive. The overwhelming cost for each individual is constant competition with every other group member. The losers

Two olive baboons face each other in a tense encounter. Every individual in a social group deals with daily threats from competitors, and the outcomes of these day-to-day interactions reinforce and advertise social rank.

may be unable to reproduce or may be excluded from access to any preferred foods or resources. Aggression within groups is normal, expected, and can be severe. Competition requires the expenditure of energy and generates personal risk. Within every primate group, there are always dominant and subordinate individuals, and a constant level of social tension exists between winners and losers.

The size and type of social groups that can be identified across the order are shaped by these opposing forces, and there are no hard and fast rules to predict the eventual outcomes. Individuals of the same species that live in areas with different ecologies may have very distinct social patterns and behaviors. Overall, the social lives of primates are a dynamic mixture of competition and cooperation. Any individual's success in terms of dominance, reproduction, and access to preferred resources may simply be a reflection of how well he or she has stabilized these social forces. Behavioral flexibility in the social domain is a key component in this equation and is clearly related to individual cognitive ability. Therefore, in addition to a number of other key factors, the ability to manage a complex social environment successfully is one of the important selective factors that has influenced the evolution of primate brains.

ARE PRIMATES MONOGAMOUS?

Monogamy is practiced by about 15 percent of primate species, including some lemurs, many New World monkeys, a few Old World monkeys, and all of the gibbons. Monogamy is also the norm for most human cultures and commonly involves an extended family composed of many monogamous pairs and their offspring. Members of this closely bonded social group frequently interact and may even live in very close association with each other. While many other primate species live in large social groups, only humans practice monogamy that sometimes involves large extended families. In nonhuman primates, monogamy generally involves groups of only one closely bonded pair of adult individuals and their offspring. All adults in these species are usually very similar in size, unlike polygamous nonhuman species, where males are generally larger than females. The age and number of offspring that remain with a monogamous pair vary among species. The offspring typically do not reproduce while they live with their parents and may provide direct care for younger siblings.

Monogamy can be considered both a mating system and a social system, evolving when the assistance of an adult male is particularly effective at ensuring the welfare of offspring. For these species, there is a direct correlation between adult male involvement and the survival of infants. Effective paternal care in primates can be either direct or indirect and is a good way of categorizing monogamous species. The smaller New World primates always have direct paternal care, whereas indirect paternal care is more common for Old World primates.

The lion tamarins (*Leontopithecus* spp.) are the classic example that illustrates direct paternal care. Female tamarins give birth to heavy babies, and twins are very common. The combined birth weight of twin baby tamarins may equal roughly 20 percent of the adult female's body weight. An equivalent human situation would be a woman of 60 kilograms giving birth to an infant of 13 kilograms. Tamarin infants are carried full time with no breaks. Without help, the physical burden on the female as she climbs, travels, forages, and carries her infants would be tremendous. Adult male tamarins actively assist in the carrying and care of the offspring, actually taking on most of the burden. Older offspring participate as well, and infants are regularly transferred among all capable members of the group. In this scenario, several factors are in play. Adult females are intolerant of any unrelated females in the group and even their own reproductively active daughters, but without assistance mothers would most likely be unable to care for their offspring properly.

Most golden lion tamarins live in extended family groupings with a monogamous breeding pair and their offspring. Unlike most primates, adult males actively assist in the care of their offspring. (Photograph by Benjamin Beck, Smithsonian National Zoological Park)

Thus, paternal responsibility becomes too great for any adult male to assist more than one female at a time effectively. In this situation, the benefits of monogamy outweigh all other costs. In some cases these monkeys practice a polyandrous social system, meaning that a group will consist of one female and two adult males, but one of these males is generally subordinate to the other and does not breed with the female when she is fertile. However, both males provide direct care to the infants. Despite the polyandrous social arrangement, it is still reasonable to call this a monogamous mating system, because only one male is reproducing.

The situation for the larger primates is somewhat different. All of the gibbons (*Hylobates* spp.) demonstrate monogamy, but none have twins or large babies. One parent can effectively carry and care for an infant. In fact, male gibbons do not provide direct care for infants, although they may actively carry and care for juveniles. Yet, the females and infants must accrue some clear benefit from the male's presence for monogamy to evolve among gibbons. Several possibilities help to explain this evolution. One is related to adult female gibbons' aggressive intolerance of any females other than their immature daughters. If males, whose essential paternal role is providing protection for females and offspring, were to breed with more than one female, they would be unable to defend multiple territories while traveling between them. For gibbons, it is less likely that protection involves predators or defending a food resource, and much more likely that it involves defending against rival males that may attempt infanticide.

As in any discussion of reproductive strategies, the interests of the males and the females must both be considered in order to understand the mating system that exists. Monogamy, like all other systems, is an adaptation that promotes the maximum benefit for the adults and the offspring while minimizing their risks.

DO ALL PRIMATES RECOGNIZE THEIR RELATIVES, AND DOES THIS INFLUENCE THEIR SOCIETY?

Like most animals, primates' behavior toward kin is different from their behavior toward unrelated members of their group. The first and most important relationship for any primate is between mother and offspring. At a very early age, infants readily distinguish their mother from all other adult females. Mothers, of course, devote special attention to their own infants. The mutual recognition between mothers and their offspring is fundamental for all primate social systems and represents a very clear understanding of a kin-based relationship.

Also a factor in these kin-based relationships is the crucial role that mothers play in the status of their young. In complex social settings where many individuals interact, a dominance hierarchy is usually well known by all members in a group. The offspring of high-ranking females enjoy more social power than those of lower-ranking females, and the basic mechanism for how this develops is straightforward. When scuffles develop between youngsters, mothers swoop in and defend their own young. Subordinate mothers defer to higher-ranked mothers and generally remove their offspring from the interaction. In effect, the dominant mother's youngster "wins" the fight because of his association with social privilege, and the "loser" takes a subordinate role. Very quickly, group members associate the social rank of the mother with her offspring. In addition to the direct influence of the mother, some primates may also make inferences about rank from watching how other individuals interact. Attention from more dominant individuals will be solicited, while group members that are perceived to be lower ranking may be challenged. As a result, the hierarchy within the group is extended to young individuals on the basis of interactions with the mother, social experiences, and perhaps observational learning as well.

These patterns of behavior persist as offspring mature. In rhesus monkeys

The offspring of high-ranking female primates enjoy fewer social restrictions than those of low-ranking individuals. Already, this young mountain gorilla is affected by the social status of its mother.

(*Macaca mulatta*), daughters usually remain in their birth group throughout their lives. The social relations and preferences exhibited by daughters have been shown to be very similar to those of their mothers, possibly transmitted by cultural learning. However, direct experiences are not the only way that primates make determinations about kin. Chimpanzees (*Pan troglodytes*) studied in captivity have demonstrated the ability to match the photos of unfamiliar adult female chimpanzees with those of their sons, indicating they are able to make judgments about kin recognition solely on the basis of physical similarities between individuals. While many animals distinguish their own kin, primates may be distinct in that they also recognize the relatives of other individuals as well.

In contrast with female primates and their unambiguous maternity, uncertainty about paternity affects the kin-based behavior of adult males. Generally, primates have either restricted or promiscuous sexual habits. Species that are monogamous or polygamous fall into the restricted category. In these cases, males make every attempt to repel other males, providing the maximum certainty regarding paternity (female fidelity is another question entirely). Adult males in these situations are highly tolerant of their offspring and in some species directly assist in their care (see *Are Primates Monogamous?* and *Do Mothers Have Help in Raising Their Babies?*). All of these fathers also provide indirect care in the form of protection. The primary threat comes from unrelated males, who are likely to attempt infanticide during aggressive encounters with the group. This behavior, which seems repugnant to many humans, makes clear evolutionary sense. If a male can remove a rival's genes from the population and perhaps have access to females that are reproductively active, there is a clear Darwinian benefit. At a minimum, this reproductive strategy relies on males' identifying unrelated offspring. As disturbing as it may be to consider, this behavior is expressed by *Homo sapiens* as well.

Far fewer species of primate have a promiscuous mating system. In these situations, when females ovulate, they may copulate with multiple males from within their group. Since paternity can never be assured, all males are usually tolerant and protective of all offspring. While it is clear that males in these social systems easily recognize in-group versus out-group individuals, their ability to assign paternity appears improbable.

The social complexities exhibited by primates have a clear relationship to kin recognition. While this appears to differ in degree among species, it is highly likely that some primates understand much more about their social relationships than most other types of animals. The bond between mothers and offspring is obvious, as well as the long-term associations that may exist among siblings. A female monkey born today may join an extended social group that includes her mother, grandmother, aunts, cousins, and sisters. Long-term studies in the wild have only begun to reveal the intricacies that exist among these relationships.

HOW DOES BODY SIZE AFFECT THE LIVES OF PRIMATES?

Primates come in a wide range of sizes, from less than 50 grams to well over 100 kilograms. Body size can be used to make general predictions about diet, locomotion, and social system.

Like all mammals, smaller primates have higher metabolic rates than larger primates and must forage efficiently for foods that meet both their energetic and nutritional needs. Insects are a convenient and nutritious package of protein and calories, although they may be a bit tricky to find. Most species of small primate rely heavily on bugs to satisfy a high percentage of their dietary needs. Some species, like tarsiers, will also consume other types of live prey such as snakes or lizards.

The diets of larger primates are based more heavily on vegetation, which has

Male and female ring-tailed lemurs are very similar in size. Both sexes contribute to group defense, and male-male competition is less severe than in species that are sexually dimorphic.

plenty of protein but is less densely packed with calories. Leaves are abundantly available and can easily satisfy the hunger of a large monkey or ape. Insects and other animal prey have the same value for larger primates but are used to supplement a vegetarian diet rather than as a staple. Fruit is highly favored by many primates, regardless of their size. It provides energy but is low in protein. Species that rely on fruit also supplement their diets with vegetation and occasionally with insects and other animal prey.

The way primates move in the trees is also associated with their size and is limited simply by the physics involved in body mass and propulsion. Small primates are able to cover gaps in the forest with leaps, using their legs to propel themselves from one tree to the next. Larger primates face a different set of circumstances. Their weight may prevent them from covering long distances in a single leap. The stability of the branch they are using may also be a factor. In order to propel themselves sufficiently, the end of a branch must be able to support both their body weight and the pushing force of their legs. As body size increases, the number of suitable branches decreases. Instead of leaping, most larger species suspend themselves below branches, grasping with their hands and/or feet. This style of locomotion allows them to support their weight safely and comfortably while moving throughout the forest.

Links also exist between body size and different social systems. In species where males commonly compete aggressively with each other, size is an advantage. Larger males are more likely to have the physical ability to dominate smaller males and subsequently increase their social status. Thus, size becomes a selective factor, inducing females to mate preferentially with larger males, who have a distinct advantage. The most desired males will be a limited commodity, and a polygamous mating system will emerge. Gorillas (*Gorilla* spp.) are an excellent example. Adult males, who may be twice the size of adult females, provide security and protection for the adult females and the offspring in their group. In species where sexual dimorphism is absent or less pronounced, males compete less directly with each other and females generally provide a comparable level of defense for the group or territory. Monogamous species such as the golden lion tamarin (*Leontopithecus rosalia*) clearly fit into this category, as well as some group-living species such as ring-tailed lemurs (*Lemur catta*). In these species, larger body size is not a selective advantage for males and does not promote reproductive or social success.

HOW LONG DO BABIES STAY WITH THEIR MOTHERS?

Pregnancy, birth, and lactation are an enormous physical drain on any female primate. The entire process of producing an infant places the mother at risk, and in some species lactating females have a higher mortality rate. After the birth, female primates must provide care and protection for the infant until it is sufficiently independent. The amount of time that a mother must devote to her offspring until weaning varies significantly among species. For some, the mother's involvement may end abruptly at weaning, while others may continue to provide other forms of care long after the youngster is capable of feeding itself.

During her lifetime, a typical female primate will be able to produce only a limited number of offspring. Each of these progeny has the potential to thrive, mature, reproduce, and therefore contribute to the mother's "fitness" (that is, the number of surviving offspring she leaves; see *Did Humans Evolve from Apes?*). Each birth is extremely important to this, and the care provided by the mother can greatly influence the chances for survival. The best way for a female to maximize her fitness is to provide the infant with nutrition and care that promote the fastest growth and weaning without placing too much physical stress on her body, so that she can re-turn to breeding condition as soon as possible. These parameters vary widely among species, but some trends are evident within the order.

The prosimians have a relatively primitive form of maternal care, with some species having litters, leaving the offspring in a nest and carrying young in their mouths. Infant prosimians develop fairly quickly compared with most monkeys and apes. The mouse-lemurs (*Microcebus* spp.) are one of the fastest, with infants being fully independent about two months after birth. For most of the prosimian species, offspring are weaned between four and eight months of age, and many are also fully independent by this time. The New World marmosets and tamarins also have ac-celerated infant development, with an intensive amount of care provided by the mother, father, and older siblings. Weaning for these monkeys may occur as early as 12 weeks after birth. The larger New World monkeys, such as the howler monkeys (*Alouatta* spp.), have a relatively greater investment in their young, nursing for ap-proximately 10 months prior to weaning.

The Old World monkeys also have infants with long periods of dependence on their mother. The yellow baboon (*Papio cynocephalus*) is an appropriate example. The infants of this species show independence around 12 months of age, and nurs-ing declines dramatically by that time. Less commonly, juveniles may be allowed to nurse well beyond 12 months, but mothers have usually become pregnant by the time their offspring is around 2 years old. While there is certainly a range within the Old World monkeys, weaning that occurs around one year of age is a useful rule of thumb.

Young apes have the longest period of dependence for any primate. Gibbon (*Hy-lobates* spp.) infants may nurse for close to two years and usually receive direct parental care well past that time. While all of the great apes have slow rates of re-production, orangutans (*Pongo* spp.) have one of the longest intervals between births for any mammal. Females may nurse their young for 3.5 years, and juveniles may rely on their mothers until they are 6 to 7 years old. Births may be spaced by up to eight years, meaning that an average female will be fortunate to have three babies in her lifetime.

Across the primate order, a trend exists between birth intervals and the relative complexity of lifestyle that a species leads. Essentially, a long period of dependence implies that offspring must learn many of the skills they need to survive over a long period of time rather than depending primarily on instinct. Female great apes in-vest years of care in each offspring, providing an opportunity for essential abilities to develop appropriately before independence is achieved. Rather than producing many offspring with a poor chance of survival, great apes produce a small number

Orangutans have the slowest rate of reproduction of any land mammal, and offspring may stay with their mothers for up to six or seven years. Birth intervals may be as long as eight years, compounding the effect that poaching and deforestation have on wild populations of this severely endangered great ape.

that are supported by adults until they are behaviorally mature. For female great apes, this strategy provides the best chance that her genes may be represented in the next generation and beyond.

DO MOTHERS HAVE HELP IN RAISING THEIR BABIES?

With the exception of some monogamous-living primates, the largest percentage of the care that young individuals receive comes directly from their mothers. In many of these species, other group members make important contributions to the well-being of infants and juveniles and assist in their physical and social development. Across the order, there is a wide range of styles related to infant care. At one extreme are adult females that spend most of their time alone and essentially have

This infant chimpanzee displays physical characteristics that are universally appealing. For nearly all young primates, a big head and large eyes are common features that evoke an immediate sympathetic response from adults. Infant chimpanzees also have light-colored skin that will darken as they mature.

complete responsibility for the care and safety of their infants. Orangutans (*Pongo* spp.) perfectly illustrate this somewhat uncommon style of primate motherhood. At the other end of the spectrum are the monogamous species in which infants may receive an extensive amount of direct care from their father and older siblings, returning to the mother only to nurse (see *Are Primates Monogamous?*). Golden lion tamarins (*Leontopithecus rosalia*) provide the example of this style of care, which also occurs in a minority of other species. By far, most species fall somewhere in between, and all primates are remarkably tolerant of the infants in their group.

Like many other types of animals, infant and juvenile primates have physical characteristics that adults find exceptionally appealing. These neotenous traits, such as a big head with large eyes, pudgy bodies, and distinct coloration, elicit a gentle response. For example, infant mantled guereza colobus monkeys (*Colobus guereza*) are born with snow white hair and pink skin, strikingly different from the dark-skinned adults, who have both black and white hair. Chimpanzee (*Pan troglodytes*) infants have lighter facial skin than adults and also have a prominent white tuft of hair on their rear end that signals their special status. Until these immature physical traits fade, infants and young juveniles are generally exempt from the group's normal social rules.

During this time frame, adult males in many species provide an important form of socialization for offspring. Youngsters commonly approach their group's alpha male without fear and may spend extended periods interacting with him. On occasion, young individuals may even engage a dominant male in a playful wrestling bout (a transgression that older members of the group would never attempt). These interactions promote an attachment between juveniles and adult males, as well as providing a model of paternal behavior toward youngsters. In some species, male tolerance toward offspring may have a practical benefit as well. Male Barbary macaques (*Macaca sylvanus*), a polygamous species, regularly allow youngsters to ride on their backs as the group travels. While freeing the mother from carrying her offspring, the male also gains an advantage. Male Barbary macaques receive significantly less aggression from other males while they hold or carry an infant or juvenile.

In addition to the direct interactions that occur between adult males and offspring, indirect support is provided to females as they care for their young. Males are primarily responsible for vigilance and defense of the group, allowing females to focus on foraging and attending to their babies. Threats come from predators but also from outside males that may attempt to assume leadership in a group. If able to enter the group, these rivals usually commit infanticide in order to reduce the number of offspring left by another male and simultaneously accelerate the opportunity to impregnate resident females. Therefore, the presence of an alpha male that is invested in the offspring of the group helps to ensure both the welfare and safety of the infants.

Infants and juveniles in a social group also benefit significantly from the attention given to them by females other than their mother. Neotenous features are a magnet for attracting female interest. "Aunting" occurs in a number species but is not particularly widespread. This behavior refers to a social group where many females collectively participate in the direct care of offspring, mantled colobus being a perfect case in point. In species that exhibit aunting, females literally compete with each other to hold and carry infants, although the mother always takes priority when nursing is necessary. Infants in these situations are surrounded by females that are anxious to provide attention and care. Young and inexperienced females benefit as well from the opportunity to refine their caregiving skills. It is not uncommon to see a young adolescent female clutch an infant to her breast but unfortunately with the bottom up and the head down. Gradually, she will understand the proper orientation and learn to avoid the same mistake when she produces her own infant.

As offspring grow, their neotenous features are inevitably replaced with a more

The dramatic coloration of this infant leaf monkey exempts it from all normal social rules and simultaneously attracts nurturing behavior from adults. As the color of the hair gradually changes with age, adults will become less tolerant of inappropriate behavior.

mature appearance. This also signals a general shift in how they are perceived by the rest of the group, ending the least restrictive period of their life. Older juveniles are forced to adhere to the social rules of the group and no longer receive the level of tolerance they once enjoyed. As they begin to accept greater social responsibilities, the care they received early in life will have helped to prepare them to be well-adjusted and confident adults.

DO MONKEYS AND APES EXPERIENCE ADOLESCENCE?

Adolescence, the developmental period when juveniles become sexually mature, is a normal stage of life for all primates. For humans, this happens during the teenage years, but in other primates with shorter life spans, it may occur at an earlier age. The process of reaching sexual maturity is generally completed before physical maturity is attained. During this transition, human and nonhuman adolescents commonly engage in some form of conflict with the mature members of their group. These social difficulties actually perform an important function, since adolescence is the time of life when primates begin the process of acquiring adult status, which may involve leaving their birth group.

For all primates, emigration consistently occurs around the onset of sexual maturity, although the specifics vary among social systems. In monogamous-living species, both male and female offspring are forced to leave, find a mate, and estab-

This adolescent male olive baboon has become peripheralized from his birth group. During this risky phase of life, he will spend greater amounts of time alone in his attempts to join another group and attract potential mates.

lish their own territory. Group-living species exhibit far greater variation, but adolescent females commonly remain in their natal group while males are ejected.

Adolescent males that leave their birth group and begin the transition to adult life face great risks, and their mortality rates are significantly higher than they are for female peers. During this time, individual males generally begin by traveling on the periphery of their former group and then venture out on their own. Without the vigilance and protection of a group, these males have a much greater risk of being attacked by predators. As they attempt contact with a new group, confrontations with more dominant males (and females) are inevitable, and these aggressive encounters pose a significant risk. Over time, some of these young males will successfully assimilate into a new group and gain sufficient status to breed with females. These fortunate individuals will have the potential to spend a portion of their adult lives enjoying social status and reproductive success. In addition to these important benefits, however, they will face constant threats from rivals who will attempt to depose and replace them.

For all primates, the adolescent transition into maturity is a combination of risk and potential success. For males in particular, adolescence is the stage of life in which natural selection can exert an especially strong influence. Those individuals that have a physical, behavioral, or cognitive advantage will be positioned to assume alpha roles in their social environment and will pass their genes to the next generation. Inevitably, even the most successful males will be deposed as they age and weaken, providing an opportunity for the next generation of status seekers.

WHY DO SOME PRIMATES HAVE SWOLLEN REARS?

One of the most eye-catching physical features associated with many primates is a brightly colored, swollen rear end. This characteristic is found in sexually mature females in a number of the Old World monkey species as well as in some of the apes. None of the prosimians or New World monkeys exhibit this trait.

The appearance of these swellings varies markedly, and depending on the species, different portions of the female's anogenital region are involved. The area that swells is called the sexual skin and may or may not include the ischial callosities. Although each species has its own characteristic type of tumescence, all are controlled by the same hormones that regulate a female's reproductive cycle. Most of the species that exhibit tumescence have a monthly cycle. At menstruation, the female's sexual skin typically has little or no swelling and lacks strong coloration. It remains this way for the first few days and then begins slowly to swell and brighten. Peak swelling and the brightest coloration occur midpoint in the cycle. Ovulation occurs

Female chimpanzees, like females among some other primate species, conspicuously advertise their reproductive status. The "sexual skin" becomes much larger and more brightly colored as ovulation approaches.

either synchronously or just after maximum tumescence and brightest coloration. After ovulation, the swelling gradually disappears as menstruation approaches.

Most of the species in which tumescence occurs have a very competitive mating system. The bright pink or reddish swelling is a strong visual stimulant for males, acting as an invitation for copulation. It is no accident that this physical feature is remarkably conspicuous, since its purpose is to attract attention. Females with a large, brightly colored sexual skin are simply irresistible to males. For females, the swelling of the sexual skin advertises fertility and reproductive vigor while drawing the maximum amount of attention. Tumescent females also become very sexually receptive and may actively solicit copulations. These occur with great frequency when a female is maximally swollen, and male competition for access to the female may be intense. Dominant males may guard ovulating females from other males until the females' swelling begins to subside. Maximum tumescence produces strong male-male competition that allows the females to mate with the most successful of the male contestants. Males also derive benefit from knowing the appropriate time to compete for access to a female, increasing the chance of reproduction.

WHEN DO PRIMATES MATE?

Different species of primate exhibit a range of reproductive cycles that are associated with a female's ovulation. The shortest of these occurs about once a month and the longest is seasonal, occurring only once a year. Frequent reproductive cycles are common for many species of primate. For example, female humans and the other great apes ovulate about 12 times in a year. Some species, such as the ruffed lemurs (*Varecia* spp.) are seasonal breeders, with ovulation lasting only one to three days per year. The males in seasonally breeding species generally also have a dramatic increase in their rate of sperm production that coincides with female ovulation.

For the overwhelming majority of primate species, mating behavior occurs only when a female is ovulating and can become pregnant. Gorilla (*Gorilla* spp.) females, who cycle about every 30 days, represent a fairly typical pattern. Each female may be willing to copulate during one or two days of her cycle but have no sexual activity at other times. Therefore, in a group with several females, the breeding male may have the opportunity to copulate fairly often. By contrast, sexual behavior for ruffed lemurs may be limited to a very short portion of the entire year, sometimes a total of only 8 to 12 hours of peak interest. In these species, as in all other types of primate, social behaviors such as mounting, thrusting, or inspecting the genitals are ordinary and occur in a number of different contexts. While not necessarily sexual, they may simply be an expression of dominance or affiliation.

Primates are able to experience much pleasure from sexual intercourse, and orgasm occurs for both males and females. However, the number of species that copulate during the infertile portion of a female's cycle is very small. While not the only examples, humans (*Homo sapiens*), orangutans (*Pongo* spp.), and bonobos (*Pan paniscus*) provide the clearest illustration of species that engage in intercourse during all phases of a female's cycle. In fact, daily copulations by members of these three species are widespread. Despite perceptions to the contrary, orangutans and bonobos regularly have intercourse face-to-face, a behavior frequently described as uniquely human (and also documented for gorillas). Orangutans, bonobos, and humans all show a range of creativity in their sexual behavior, engaging in activities that are based on pleasure rather than reproduction, such as oral sex. Same-sex erotic interactions are the norm for a segment of the human population and are also practiced by orangutans and bonobos, both in the wild and in captivity.

Sexual interaction is always a fundamental aspect in the social behavior of any species. When reproduction and sex are not necessarily tied together, novel social opportunities emerge. The pleasure derived from sex may be a means to exert influence, settle disputes, or promote a relationship. In addition to reproduction, sex becomes one of the fundamental aspects related to social politics.

DO MALES OR FEMALES INITIATE MATING?

Securing the opportunity to mate and potentially reproduce is a fundamental force that shapes the behavior of individual primates as well as their social systems. Both males and females are driven to pass their genes to the next generation, but they approach the challenge with reproductive strategies that may be in direct conflict. Over the course of her lifetime, a female's biology limits the number of opportunities she has to procreate. Gestation, lactation, and the rearing of her offspring are time and energy consuming and may place her at considerable personal risk. By comparison, males have almost no biological limit related to reproduction, and fertilization is an energetically cheap investment. Although in some species males provide direct paternal care for their offspring, most protect their progeny through indirect means, such as vigilance and group defense. These undeniable differences create conflict between the sexes regarding reproduction. For a female, the survival of each of her offspring carries great value, and her choice of a mate has a pronounced influence on the potential success of her offspring. Males may best increase their fitness (see *Did Humans Evolve from Apes?*) by focusing less on the quality of their mates and more on the production of large numbers of young. Not surprisingly then, reproductive strategies related to mate choice differ between males and females.

Healthy males that demonstrate their physical confidence have a clear advantage with females. In addition to the basic desirability of vigor, stronger males are more likely to outcompete other males and gain social status. Capuchin monkeys (*Cebus apella*) and vervet monkeys (*Chlorocebus* spp.) are species where females clearly prefer high-ranking males as their reproductive partners. This is not true for all species or in all circumstances. Incest avoidance may prevent females from mating with high-ranked males that are relatives. In these situations, lower-ranking nonkin males are clearly preferred. Additionally, social rank may also be less influential than social behavior in female mate choice (discussed further below). Males that are able to acquire status through physical prowess may rely on behavioral strategies that prevent females from consorting with other males. Alpha males may simply guard receptive females, making themselves the only available mate. In some species, such as chimpanzees (*Pan troglodytes*), males may harass females until they allow copulation. Using a strategy that is very uncommon for primates, orangutan (*Pongo* spp.) males will pursue and forcibly copulate with females. Among males in some promiscuous species, such as many of the macaques (*Macaca* spp.), an additional feature protects their genetic investment; after ejaculation, the semen of these males creates a plug that lodges in the female's reproductive canal. The first male to copulate with a receptive female has a clear benefit. Although other males may subsequently deposit sperm, access to the ripe ovum will be blocked.

The largest and strongest males are not always the most reproductively successful. Younger males, such as the western lowland gorilla shown here, may be preferred by females if they demonstrate superior social skills.

The first and most important aspect related to female mate choice is the recognition of a desirable male. In species that live in groups, females use appearance and behavior to assess the quality of a potential mate. Both of these indicators can have a significant impact on her choice. Contrary to traditional explanations, the largest and most dominant males are not always the individuals that are most preferred. Females may be attracted to males that demonstrate particularly strong social skills, such as those that are most solicitous toward their young offspring. Once a reproductive partner has been identified, females will demonstrate obvious receptive behaviors toward that male and may openly invite copulation. Females among western gorillas (*Gorilla gorilla*) and chimpanzees (*Pan troglodytes*) can be particularly persistent, making their intentions obvious by following a male very closely and staring at him. As a final form of persuasion, female gorillas have been known to grab and stimulate a male's genitals in order to initiate copulation. Within a social group, mate choice may also be accomplished by rejecting an unpreferred male's sexual advances. Females may turn their back, avoid spending time around certain males, or simply refuse to copulate. On occasion, females will also aggressively respond to an unwelcome advance and may rely on the assistance of other females to drive off an unpreferred male.

Transfer between groups can be considered an obvious expression of female mate choice. In some social structures, females avoid mating with related males by leaving their natal group and integrating themselves with nonkin. These situations are easy to understand and promote healthy genetic diversity in a population. In a secondary transfer, a female leaves the group in which she had been living, along with its resident male, clearly demonstrating mate choice (and mate rejection). Female mountain gorillas (of the subspecies *Gorilla beringei beringei*) rarely have access to outside males, except during conflicts between groups. Secondary transfers occur most often by females during these intergroup encounters, and males may attempt to prevent females from leaving. However, males with few or no female companions may initiate contact with other groups in an attempt to attract additional females. Females that have dependent offspring do not move between groups.

A final reproductive strategy that is used by females involves maintaining their place within a social group but engaging in extragroup mating with an unrelated male. These consortions have been documented in chimpanzees (*Pan troglodytes*), a species where females sometimes travel alone and normally spend time away from other members of the group. Clearly, these matings occur away from the group's males, who could dominate these females and attempt to prevent such matings. With the advent of noninvasive means for assessing paternity, such as DNA analysis from hair or fecal samples, the extent to which these extragroup matings occur in wild primate populations has only recently begun to be understood.

While some male and female reproductive strategies may be obvious for group-living species, much remains to be learned. The roles that personality, temperament, individual preference, and pheromones play are open for investigation. Monogamous species have the most obvious pairings, yet the mate choice mechanisms involved in these relationships have not been thoroughly studied and are not well understood.

WILL PRIMATES ADOPT A BABY?

Adoption is a common behavior for humans and has also been documented to occur in a number of other primate species as well. Adoption always implies that a self-sufficient individual voluntarily assumes responsibility for an infant or juvenile who is not capable of caring for itself. This added responsibility is time and energy consuming and is most easily understood when a relative is involved. For example, a brother and a sister share some genes with each other as well as with each other's offspring. Adopting a sibling's offspring ensures that the family line will persist and that those shared genes will be represented in the next generation. Adoption of an unrelated individual cannot be explained in the same way and suggests that the caregiver's actions may have altruistic motivations. As with humans, adoptions in other species of primate are most likely performed by adult females but have also been accomplished by adult males and siblings. Jane Goodall reports that wild orphan chimpanzees in Gombe, Tanzania, have been adopted by both male and female older siblings, who provide equivalent levels of care. Adoptions by unrelated individuals have also been documented.

Clear examples of kin-based adoptions have occurred both in captivity and in the wild. Fossey, a 14-month-old gorilla (*Gorilla gorilla*) living at the Columbus Zoo in Ohio, was adopted by his father, Bongo, upon the death of his mother. Bongo, who was just over 30 years old at the time, immediately became the full-time caregiver for Fossey. He allowed the baby to sleep with him and to ride by holding on to Bongo's arm as he walked, and he refused to eat any of his daily diet until Fossey had taken all of the preferred items first. Like many adult male gorillas, Bongo did not regularly make a night nest for himself, preferring to lie directly on the ground while sleeping. Once he adopted Fossey, Bongo was observed to make a very small nest against his body in which the baby slept during the night.

Great apes have also been observed to adopt their younger siblings. Indah, an adolescent female orangutan (*Pongo* spp.) at the National Zoo in Washington, D.C., became a surrogate mother for her three-year-old sister, Iris. In all ways, Indah assumed the full care of her sibling, sharing food and a nest and even allowing her to nurse. The stimulation from Iris's suckling caused Indah to begin to lactate, even though she had never been pregnant.

The only known adoption of an infant under three years of age by a mountain gorilla (subspecies *Gorilla beringei beringei*) occurred in the Virunga volcanoes of Rwanda in 2002. An adult female had been killed by poachers, although her 13-month-old infant survived and clung to her body. This very young female was rescued by the staff of the park and quickly regained her health after medical treatment. Taken back to her group, she was instantly surrounded by its members upon her release. The gorillas appeared quite agitated and became relaxed only after the dominant male in the group approached, inspected her, and then calmly moved away. Despite the presence of many lactating females with infants, none attempted an adoption. However, this young female had a 10-year-old brother in the group, and he apparently recognized his sister, adopted her, and immediately assumed the role of her caregiver. This very young female appeared healthy after the adoption, although her long-term survival cannot yet be guaranteed. On cool evenings, she

Mandara, a western lowland gorilla at the Smithsonian's National Zoo, spontaneously adopted an infant that was rejected by another female in her group. Here, Mandara cradles the newborn while nursing her own one-year-old male offspring. (Photograph by Jessie Cohen, Smithsonian National Zoological Park)

has been seen sleeping tucked between her brother and father, the dominant male in the group.

At the National Zoo, an exceptional adoption event occurred involving four gorillas (*Gorilla gorilla*). Holoko, an adult female, gave birth to a male infant named Baraka. For unknown reasons, she demonstrated no interest in this healthy infant and immediately abandoned him. At that moment, Mandara, another adult female in the group who was caring for her own 12-month-old infant, Kejana, approached Baraka and instantly adopted him. Mandara then cared for both Baraka and Kejana, even nursing them simultaneously. Within a few months, however, Kejana slowly began to seek out Holoko's company, most likely because of his own mother's divided attention. Holoko spent more and more time with Kejana and eventually adopted him as her own. At that point, the two females were providing full-time care for each other's offspring. Most important, neither was related in any way to the youngster that she held, carried, comforted, and treated as her own.

HOW HAVE IDEAS ABOUT PRIMATES CHANGED?

Reports from explorers about wild primates have existed for hundreds of years, the earliest of these occurring over 3,000 years ago. These reports document humans' awareness of primates outside their native countries well before the seventeenth century. Although these records were considered and discussed by academics and philosophers, the details from them are unfortunately quite vague, and the surviving information largely resembles myth and legend.

In the 1600s, reports began to contain enough detail to allow for identification of the species that were being encountered in the wild, and it is almost certain that orangutans and chimpanzees were involved. The characterizations that were applied to these great apes reflected the values of the times. Orangutans were considered modest; females were reported to "hide their secret parts" when being observed (Yerkes and Yerkes 1929, 12). Male orangutans were reported to kill men in order to steal their women, whom they kept for "the pleasure of their company" (Zuckerman 1963, 8). These writings also described the social behavior of large groups of monkeys, who were thought to live in a highly organized society run by chiefs and assumed to be simply too lazy to build homes for themselves or grow crops.

Primate studies in the 1700s focused heavily on classification of the order, and interest in behavior and natural history became more obvious during the 1800s. In this period, primates were understood purely in human terms. Some scientists asserted that the behaviors, abilities, and even the appearance of apes were simply underdeveloped versions of what was exhibited by humans. In this sense, apes were characterized as only a "rough model" of humans, who were considered to be more highly evolved and "polished." Early descriptions of primate social systems began to emerge in scientific writings and, although preliminary, contained some details that are still considered accurate today. Despite this accurate information, the period's illustrations of wild apes usually depicted them as a traditional family, with a female clutching an infant to her breast while being protected by an adult male.

Throughout the 1900s, behavioral studies of primates both in the wild and in captivity gained prominence and scientific rigor. In the first half of the century, notions about primates as savage and barbaric began to be discarded. Perhaps as compensation, romanticized views gained favor. The social behavior of wild apes was characterized as overly peaceful and tolerant. Gorillas are one example where unrelated adult males were thought to accept and interact with each other freely, no territories were defended, and rival groups met and happily joined together. Since that time, fieldwork has revealed that the truth is somewhere in the middle. Apes are not savage, nor do they live in the Garden of Eden. Aggression and peacefulness are normal facets of primate social systems and vary depending on individuals, circumstances, and social pressures. Great apes in particular are capable of acts that many would considerate compassionate or even moral. At the same time, cruelty and violence have also been documented. Primate social systems are complex, dynamic, and clearly much more than a distilled version of human society.

The study of primate social systems has gradually moved away from strictly anthropocentric notions and now focuses on understanding different species in their own right as well as on attempting to identify both the similarities and differences that exist between human and nonhuman primates. As in any field of science, progress depends on collaboration and incremental steps forward on the basis of existing work. Today, building on the foundation established by the Japanese school of primatology in the mid-1900s, scientists are devoted to the study of social learning, innovation, and culture among primates. These topics are vital to a more complete understanding of primate social behavior and have established a firm research agenda to begin the twenty-first century.

WHY DO PRIMATES SPEND SO MUCH TIME GROOMING?

All primates engage in grooming. While this is one of the most commonly observed primate behaviors ("Are they picking fleas?"), it is also one of the most misunderstood. There are two types of grooming: autogrooming and allogrooming. Autogrooming refers to an individual's self-grooming. For example, an adult male chimpanzee might spend a part of the day examining his body in detail, removing any bits of dirt or dry skin that he finds. He might also use a small twig to probe and clean a wound or scab. If any parasites are found, they would be immediately removed and discarded. Autogrooming primarily serves a hygienic function.

Allogrooming refers to the grooming that one individual gives to another. In this case, grooming involves the slow and methodical process of one individual touching, inspecting, and cleaning any part of another individual's body. Allogrooming bouts may be short in duration or may occupy an entire afternoon. During an extended grooming session, the groomer closely attends to the tiniest details of hair and skin, painstakingly parting hair, cleaning and focusing on his or her job. The groomee accepts this affection openly. Commonly, the individual being groomed will achieve such a state of relaxation that he or she struggles to stay awake. While being groomed, the expression on the face speaks of pure delight.

Allogrooming may involve a mother and an infant or any possible combination of juveniles and adults. While hygiene was almost certainly the origin of allogrooming, it is probably the least important aspect of the behavior that we see today. Within a primate group, allogrooming creates and reinforces social bonds

Chimpanzees, like all other primates, engage in autogrooming. This behavior is related specifically to personal hygiene rather than social interaction.

among individuals, eases tensions after a confrontation, advertises interest to potential sexual partners, and, among many other functions, can simply express affection between friends. This kind of grooming can be thought of as social currency among primates.

Allogrooming is an invaluable way to assess the status of each individual within a social group. The amount of grooming that an individual receives is a reliable indicator of rank. The more central individuals are to their group, the more likely it is that they will receive more grooming than they give. Over time, the relationships that each group member forms, maintains, or abandons can be tracked in respect to his or her grooming habits. Typically, two individuals that have a very close bond, such as two mature females, groom each other reciprocally on a daily basis. If problems emerge in their relationship, grooming patterns may change. Bickering may produce an increase in their grooming rates, indicating a need for tension release. If grooming diminishes or stops completely, the social distance between them may be permanent.

In some primate species such as macaques, baboons, and vervets, stages of infant development can also be correlated with grooming rates. Initially, infants are

All primates, such as these chimpanzees, reveal their social relationships through allogrooming. The way in which individuals pair off during grooming bouts can indicate rank, alliances, or simply affection between friends.

groomed intensely by their mothers, sometimes while nursing. While grooming is always obvious between mothers and offspring, it decreases as the young individual becomes more independent. Young primates, especially females, demonstrate social development when they begin to groom other group members. Once young females mature, the rate of reciprocal grooming with their mother is consistently high.

Grooming by males follows very different patterns. Young males spend less time grooming their mother as youngsters. In fact, grooming between mothers and sons never becomes reciprocal. Over time, young males may be excluded from the grooming that occurs in female social circles. At this stage of development, young males begin the process of leaving their birth group. Entry into a neighboring group is difficult but can sometimes be facilitated by approaching and grooming low-ranking females, a typical male strategy. Males also demonstrate their interest in sexually receptive females by offering to groom.

While clear gender differences are evident in grooming behavior, the importance of grooming itself is not. Both males and females rely on grooming to express their loyalty, interest, affection, and social bonds with other group members. On a larger scale, grooming facilitates group cohesion, conflict resolution, mating preferences, and familial bonds. This simple activity is one of the behaviors that demonstrate the remarkable social complexity that exists for most primate species.

WHICH PRIMATES HUNT FOR MEAT?

Tarsiers (*Tarsius* spp.) are highly carnivorous, tamarins (*Leontopithecus* spp.) regularly eat small animals such as frogs, and macaques (*Macaca* spp.) consume a variety of prey such as crabs and even fish. Meat eating is not particularly uncommon among primates, but prey is usually found during general foraging activities when other types of food may be eaten as well. Hunting is distinct from foraging, involving the search and pursuit of a specific type of prey while other types of food are ignored. Of course, hunting is well known among humans (*Homo sapiens*) and is presumed to have had a major influence on the development of the social and mating systems of human ancestors.

Until the late 1960s, it was thought that humans were the only type of primate that engaged in hunting behavior. Since then, both olive baboons (*Papio anubis*) and chimpanzees (*Pan troglodytes*) have been documented to hunt for prey. Baboons consume a number of different types of meat, such as hare and small antelope, and most of these appear to be found serendipitously. Pursuing and eating animal prey is a predominantly male activity for baboons. On the rare occasions when a female successfully catches a small animal, a more dominant male usually takes it from her. Baboons have been reported to visit and search areas where prey is likely to be found, strongly suggesting that they are actively hunting for meat. Baboons do not cooperate in the capture of prey, nor do they share meat with each other. Hunting by baboons appears to be an activity pursued by individuals, and meat eating is primarily reflective of acquiring a highly favored type of food.

Meat eating by chimpanzees encompasses a more complex array of behaviors as well as a much broader range of cognitive complexity. The types of prey consumed by these great apes include young bushpig, antelope, and terrestrial monkey. These usually appear to be found serendipitously and are then pursued and eaten if caught. However, if an adult bushpig is seen, the chimpanzees will actively search the immediate area for a nest that may contain piglets. By far, the most common and preferred prey are the tree-living colobus monkeys (*Colobus* spp.), which constitute nearly 85 percent of the meat eaten by some populations of chimpanzees.

The highly arboreal lifestyle of colobus monkeys necessitates a capture strategy very different from that used for terrestrial prey. As with humans and baboons, hunting by chimpanzees is largely a male activity. The general tactics used by chimpanzees to capture colobus monkeys vary among populations and represent clear cultural differences. In some cases, chimpanzees successfully hunt alone, or multiple individuals will initiate a hunt if colobus monkeys are encountered in the forest. In other situations, groups form hunting bands and set out in search of this favored prey. These groups move through the forest, scanning the treetops. Once monkeys are located, the hunters move into position. In some populations, it appears that the behavior of the group is not coordinated and that they all attempt to capture prey simultaneously. Other hunting parties behave very differently, and each member of the band takes a specific role. These coordinated hunts have been reported to involve "blockers," which prevent the monkeys from escaping while "drivers" move them toward chimpanzees that have concealed themselves in the treetops, waiting to capture their prey.

After a successful hunt, chimpanzees routinely share the meat with each other. As in all other areas of chimpanzee behavior, the distribution of this highly preferred resource involves politics. Meat may be given in order to solidify social alliances or denied in order to reject rivals. In certain populations, the single best predictor of when males will hunt for meat is the presence of an ovulating female in their group. After a successful hunt, males share meat with swollen females, which some researchers believe may be a direct exchange for sex. If this interpretation is correct, this behavior would demonstrate a strategy in which males increase their likelihood of reproducing, while females gain a highly preferred food and copulate with a male that has participated in a successful hunt. For chimpanzees (and most likely ancestral humans as well), hunting represents more than simply a way of obtaining a preferred food. The meat that is obtained gives the hunters a means of maintaining their social status as well as a way of increasing their reproductive potential by enticing copulations from receptive females.

DO ALL PRIMATES HAVE FRIENDS?

Any discussion of primate social behavior is likely to include references to "leaders," "rivals," "winners," and "losers." The effective use of these labels is not based on a hunch but rather on studying the interactions among individuals and the outcome of competitions. These terms are used commonly and without hesitation to characterize members of a group. In contrast, words such as *friends* were historically considered overly anthropomorphic and therefore avoided. Now, although caution against overinterpretation is always prudent, *friends* is considered a perfectly appropriate term as long as its use is based on observations of behavior.

Friendship among primates is distinct from other relationships in a group. Friends spend more time in close proximity to each other than to others in the group and groom each other much more frequently. These close bonds may exist among females, males, or males and females. Friendship is also a good way to predict kinship, support during fights, and even sexual behavior.

Friendship has been particularly well studied in olive baboons (*Papio anubis*), Barbary macaques (*Macaca sylvanus*), and rhesus monkeys (*Macaca mulatta*). In these species, females may form very long-lasting and strong friendships with males. Usually, females become friends with a male who is already a friend of their female relatives. A male's friendship with a female extends to her offspring as well.

A female acquires clear benefits from her male friend. He provides her and her offspring with protection from aggression from other males and also from any interference while she is feeding. The most obvious benefit for a male is that he may be a preferred sexual partner when a female friend is ovulating. At other times, females provide important support to their male friends during aggressive encounters with rival males.

The ways in which a friendship is maintained between a male and a female are indicative of the benefits that each receives. Since females profit from the presence of a male friend, they spend more time initiating close proximity and also provide a greater share of the grooming that occurs between them. During uncomfortable encounters with rivals, males seek out their female friends for support. In olive baboons, a female with a small infant will spend more time with a male friend in a clear attempt to avoid aggression and potential injury to her offspring. Although most likely unrecognized by either of the baboons, this male friend is commonly the father of the infant.

A female olive baboon expresses her friendship with an adult male through grooming. The presence of an adult male reduces the risk of predation or aggression from other baboons for both the mother and infant.

Clearly then, in addition to having enemies and competitors, primates also have friends. Some of these friendships fall apart and are abandoned, whereas others may persist for many years. All close friendships develop gradually over time, and each individual makes an effort to maintain the relationship. Tangible benefits, such as protection from aggression, can be measured and documented. The intangibles that friendship provides are less obvious and can be measured only by the warmth and comfort shared between two good friends, whether baboon, human, or macaque.

DO PRIMATES "MAKE UP" AFTER A FIGHT?

One of the most interesting elements related to the balance between conflict and cooperation is reconciliation, the ability to restore an existing friendly relationship after a clash between individuals. Conflict and cooperation are basic elements in any social group of primates. There are no examples of a social system that has only one of these features but not the other. The level of cooperation and conflict that exists in any collection of individuals can vary significantly depending on a number of factors such as species, circumstance, mating system, and social system.

Reconciliation usually occurs with a predictable pattern. At some point after a fight, one individual will approach the other and attempt a peaceful gesture. The timing of this act must be considered carefully. If the attempt at reconciliation is too soon (or too late), aggression may reerupt and cause further damage to the relationship. If the overture is accepted, the process of repairing and restoring the social bond can begin. For primates, the most common form of explicit reconciliation involves physical contact. Depending on the species or the degree of conflict, reconciliation can involve a range of behaviors including a gentle touch, an embrace, intense grooming, or vigorous sexual activity.

Reconciliation does not occur with equal frequency in all groups or between all individuals. For example, stump-tailed macaques (*Macaca arctoides*) have very frequent, obvious reconciliations, whereas rhesus monkeys (*Macaca mulatta*) are much more subtle in their attempts. One of the central factors that can influence an individual's desire to reconcile is the importance of the relationship that has been affected. Relationships that are highly valued involve a shared investment by both parties, and therefore the desire for repair should be mutual. In general, the most important relationships for primates are ones that involve support during aggressive encounters or direct protection from rivals. Not surprisingly, reconciliations can be best predicted to occur between individuals that provide each other with this form of safety and defense. Relationships that are based on behaviors such as group hunting, food sharing, and mating are also important but probably less influential overall. For male chimpanzees (*Pan troglodytes*), alliances with other adult males are absolutely crucial in their social system and are highly prized. Therefore, male chimpanzees are strongly motivated to reconcile with each other after a fight and do so frequently. By contrast, adult male orangutans (*Pongo* spp.) never form alliances and are strongly competitive with each other. No reconciliation between adult male orangutans has ever been reported to occur. Many species of macaque (*Macaca* spp.) live in mixed social groups that have a high degree of competition among males. Predictably, the highest rates of reconciliation are between males and females in these social systems.

Conflict, cooperation, aggression, and reconciliation are all normal features that are expected in any group of primates. The relative levels of these behaviors can be descriptive of a species, since some are simply more peaceful or violent than others. However, every socially living type of primate shares the same need for a network of support. Friendships and alliances have great influence on an individual's social status. Despite perceptions to the contrary, brute strength can be far less important for gaining social rank than skill and finesse in maintaining a valued relationship. The decision to attempt a reconciliation can be strategic and may have a profound influence on the power structure within a collection of individuals. The continuation of a strong alliance can provide the foundation for an entire group, while a shift in the balance of conflict and cooperation can create a severe disruption. A single act of reconciliation may be the difference between stability or social turmoil.

DO PRIMATES MAKE GOOD PETS?

Baby primates are very appealing, and the impulse to have one for a pet is easy to understand. Pet stores and dealers that sell primates frequently focus on how cute they are and market them as adorable little "people" that will become welcome additions to any family. In reality, the personal risks that are involved for owners, as well as the inherent animal welfare concerns, should dissuade anyone from considering any type of primate as a pet.

In many areas of the United States, private ownership of primates is illegal. However, dealers in these same areas may operate legally and are not necessarily obligated to inform customers of the local laws. Normally, the primates being sold in the pet trade are infants or very young juveniles. They may still be of nursing age and require bottle-feeding, which unfortunately adds to their appeal for many potential buyers. At any stage of life, humans and all other primates can share almost every type of infectious disease. Humans can easily contract serious or life-threatening conditions from many of the primates commonly sold by dealers, even the ones that come with health "guarantees." As primates mature into adolescence and adulthood, their behavior can change dramatically. A formerly tractable juvenile can become quite aggressive, destructive, and dangerous to people. These changes are perfectly normal and reflect the typical ways in which primates develop as adults.

In addition to the genuine safety and health risks posed by pet primates, there are animal welfare concerns that simply cannot be avoided. Typically, infant primates that are sold as pets are taken from their mothers very soon after birth to encourage socialization with people. Female primates form very strong bonds with their infants that may last for a lifetime, and the effect that these forced separations have on the mother must be considered. Over the course of her reproductive years, a breeding female may have numerous infants taken from her to supply the pet trade. There can be no doubt that this severely compromises her psychological welfare and may be detrimental to her physical health as well.

Despite what may be the best of intentions, the owners of a pet primate simply

Although infants and juveniles appear adorable, the physical and behavioral changes that accompany normal primate development should discourage anyone from considering any type of primate as a pet.

cannot provide the companionship and social environment that every primate requires. Humans can never adequately replace the quality of social interaction that a member of the same species can provide. As primates become adults, a poor social environment combined with increasing strength easily creates the opportunity for aggression directed at people. At this point, exasperated owners are forced to keep their primate constantly confined, which only increases their pet's frustration. Since many primates that are commonly kept as pets can live for decades, this unfortunate cycle usually leads to a situation that is intolerable for both the pet and the owner.

Frequently, primate dealers suggest that all of these problems can be avoided with proper veterinary intervention. Many pet primates are castrated or have healthy teeth pulled in order to prevent aggression or injury to owners. These medical procedures may reduce the risk to people, but the primates are permanently compromised as a result. Proper socialization of a former pet may be impossible once they have been surgically, and therefore behaviorally, altered.

The best way to prevent all of these difficult situations is simply to avoid considering a primate as a pet. For those individuals who clearly have an enduring interest and fascination with primates, volunteer opportunities are one of the best ways to promote primate welfare and personal satisfaction. Most zoos and sanctuaries that house primates have a very active volunteer force and provide a productive and meaningful way to spend time around a variety of primate species.

3

PRIMATE INTELLIGENCE

HOW SMART ARE PRIMATES?

Smart is a subjective term, and it can mean different things when applied to various situations. For instance, all primates are well adapted to the environments in which they have evolved and are perfectly capable of making a living on the basis of what they know about those surroundings. In terms of local ecology, primates are extremely smart.

Socially, primates are fully capable of cooperating with each other, forming alliances, and communicating their intentions. They learn from each other, develop traditions, and even show cultural variation in their behavior. Some primates appear to engage in active deception and even base their actions on what they think other individuals know (see *Are Humans the Only Deceptive Primate?*). Politically, primates are definitely smart.

When their cognitive skills are tested, they demonstrate an understanding of logic, abstract thinking, and rules and contingencies. Primates are able to devise novel solutions to accomplish a specific goal. They are endlessly curious, potentially creative, frustrated by boredom, and attracted to new challenges. In terms of comparing the mental abilities of all other primates with those of *Homo sapiens*, we are only beginning to learn how smart they really are.

HOW DO SCIENTISTS STUDY PRIMATE INTELLIGENCE?

Researchers approach the study of primate intelligence from a variety of different disciplines and with a range of academic backgrounds. As in all other areas of science, progress is expected to be incremental, collaborative, and, it is hoped, self-

Chimpanzee (*Pan troglodytes*)

Many species of primate, such as this gorilla, consistently demonstrate an impressive array of mental abilities. Do they reflect on the past and consider the future? How different is the human mind from those of our closest evolutionary relatives? Current research is attempting to find out.

correcting. New information is added to the existing literature in an overall attempt to reveal facts that stand the test of time. Studies that have roots in psychology, biology, anthropology, and ecology (among other disciplines) have all made significant contributions to the study of primate intelligence. None of these areas of expertise alone is sufficient to understand primate behavior and intelligence fully, but each contributes information that helps to describe more fully the range of abilities and mental skills that are present for primates.

Most of the scientists that work with primates specialize in either fieldwork or captive studies, which can exert an influence on the specific area of cognition under investigation. Fieldwork traditionally relies on purely observational methods for collecting data. The wild primates under study have become habituated to the presence of researchers, who avoid any actions that might influence the behavior of their subjects. This style of research, conducted in a natural habitat, is ideal for considering questions about such issues as mental mapping skills and foraging habits. Food acquisition is a particularly important topic of study and may include tool-using abilities, judgments about quantity, and a variety of social behaviors such as cooperation and observational learning. In the wild, the abilities exhibited by individuals as well as by local populations in general can be documented.

Some of these topics are simply impossible or too difficult to study properly outside a natural habitat, but those that must be studied in the wild also confront scientists with difficulties and limitations. These difficulties may be as simple as finding subjects in a dense forest and attempting to follow them as they move rapidly through the trees. Visibility can be problematic at times, and predicting when certain behaviors might occur can be difficult. Perhaps the greatest difficulty associated with fieldwork is the enormous range of uncontrollable factors that can influence the behavior of the subject primates. However, even though these may have a negative effect on some investigations, they can also create situations that generate new studies.

Research conducted in captivity draws on a wider array of methodologies for exploring cognitive skills, ranging from fieldwork-style observations to direct interactions between experimenters and their subjects. Scientists that investigate cognition in captivity generally work with a small number of subjects and specifically focus on the abilities of individuals rather than populations. Studies may be short-term or longitudinal and are generally well suited for revealing ranges of cognitive capacities and styles of learning. On the practical side, captivity provides researchers with the opportunity to document the most detailed aspects of behavior and to present subjects with controlled experiments that allow for exact interpretation of results. These advantages allow scientists to appreciate aspects of certain cognitive skills that may be impossible to study in the wild, such as social learning, tool use, and communication. Of course, captivity is less appropriate for dealing with questions that concern the ways in which primates interact with their natural environment, and studies that focus on a small number of individuals may not properly represent the average level of mental ability found in a species overall.

Studies of primate intelligence in the wild and in captivity have their own individual advantages and risks. Fieldwork is particularly good for documenting the range of natural abilities that exist in a population. Captive studies can be most effective at describing the process of how individuals learn and in providing mental challenges that reveal the range of cognitive flexibility present for a particular species. The study of primate intelligence is best served by the interaction between scientists working in the field and those working in captivity. These different styles of investigation complement each other, and active collaboration advances the overall understanding of the mental skills that are present throughout the primate order.

WHEN WAS PRIMATE INTELLIGENCE FIRST STUDIED?

Most of what is known about the mental abilities of primates was discovered in the last half of the twentieth century. Prior to that, long-term fieldwork was virtually nonexistent, and captive studies were uncommon. Three notable exceptions occurred around the turn of the nineteenth century, and each made a valuable contribution to the study of primate intelligence. The research that continues today builds on the foundation that was established by early pioneers in the field of primatology.

The earliest attempt to study the mental abilities of primates systematically began near the end of the 1800s, when Richard L. Garner investigated the intelligence of monkeys. His studies largely focused on the ways in which captive monkeys communicate with each other and attempted to resolve the question of whether they have their own language. Garner relied on a blend of experiments, informal observations, and anecdotes for drawing conclusions about the abilities of monkeys. He also applied the technology of the day, using a phonograph to capture the vocalizations of his subjects. These recordings were then played back in order to document the reaction of the monkeys. This playback methodology is well accepted today and is commonly used to study the behavior of many different species.

Garner also ventured to the forests of Africa in order to study the behavior and vocalizations of great apes. His philosophy in regard to studying primates in the wild was revolutionary for the time. In an era when great apes were considered savage beasts and hunted as trophies, Garner attempted to witness their natural behaviors unobtrusively. Not completely immune to the prevailing perceptions of the day, he watched and listened from the inside of a large cage, which he had built in the middle of the forest to ensure his safety. On the basis of observations and information collected during his travels, Garner correctly reported that gorillas are polygamous and that chimpanzees live in groups of multiple males.

Despite his admiration for and fascination with the cognitive abilities of primates, Garner's conclusions about the language skills of monkeys and the mental abilities of primates were heavily influenced by speculation and anthropomorphism. For the era in which these studies were conducted, this is not unexpected. However, the innovative ways in which Garner approached the study of primates clearly illustrate that he was a man decades ahead of his time in terms of philosophy and attitude. While the substance of his research may have been flawed, his earnest desire to understand and explore the mental complexity of monkeys and apes provided legitimacy for a field of study that was only beginning to emerge.

In the early 1900s, great apes fared poorly in captivity, and few individuals survived into adulthood. One consequence of this was the rare opportunity for their study. In 1913, the Prussian Academy of Sciences established the Primate Station on Tenerife Island, off the northwest coast of Africa, in order to support the study of chimpanzees in a climate that was more conducive to their good health. The station was the first captive situation ever established for the specific purpose of studying the behavior and cognition of primates. The first director, Eugen Teuber, emphasized the importance of establishing a positive relationship with the chimpanzees at the station and provided the apes with comfortable living quarters in which they could explore, play, exercise, and socialize. When Teuber's term ended, he was replaced as director by Wolfgang Köhler. While Teuber is rarely mentioned in connection with the Tenerife station, Köhler is well known for his groundbreaking studies of chimpanzee cognition.

Köhler chose to work with chimpanzees for two general reasons. The first was practical, associated with the physical and behavioral resemblances between chimpanzees and humans. Köhler recognized clear manifestations of these similarities in the ways that chimpanzees appear humanlike as they carry out their normal, everyday activities. The second reason was theoretical, based on the assumptions that chimpanzees possess a more original form of intelligence than that of humans and that the study of their mental abilities would lead to a greater understanding of the intelligent behavior exhibited by humans.

Köhler was specifically interested in documenting "insight" rather than more traditional forms of learning, such as trial and error. He presented his nine chimpanzee subjects with a variety of experimental situations that could be solved through insight, but he provided no instruction or guidance related to potential solutions. His most famous experiments presented the chimpanzees with out-of-reach foods that could be obtained only with the aid of a tool. The chimpanzees demonstrated a variety of different techniques for solving the problems. For foods that were suspended overhead, the apes moved boxes underneath and stacked up to three together in order to obtain the reward. In some cases, one chimpanzee would steady the boxes while another climbed to the top. Other individuals stood a tall pole underneath the food, rapidly climbing and successfully grabbing the reward as the pole fell over. If food was placed beyond reach outside their enclosures, the chimpanzees constructed and used reaching tools. The most elaborate of these was a combination of three different hollow sticks that were joined end to end to make one long tool.

Many of Köhler's most interesting observations were made in situations where no specific task was presented to the apes. On a number of occasions, for example, the chimpanzees attracted chickens to the fence of their enclosure by offering them small bits of bread. Once within range, the apes stabbed them with sharp sticks or pieces of wire. Long before chimpanzees were studied in the wild, Köhler documented many of their natural behaviors such as nest building, digging with sticks, and using a tool to fish for ants, which were readily consumed.

Köhler's early contributions to primatology were profound. His carefully designed studies and detailed observations provided the first experimental evidence to support the idea that nonhuman primates are capable of tool use, problem solving, planning, and insight. At almost the same time as Köhler's classic studies were in progress, the study of great ape behavior and cognition began to emerge independently as a major field in the United States.

Robert M. Yerkes has the distinction of being the individual who is most responsible for establishing the academic foundation for primatology in the United States. Around 1916, Yerkes began his research on the behavior and cognition of great apes. His studies had a much broader scope than those of Köhler and included gibbons, orangutans, chimpanzees, and gorillas. At the time, bonobos were not recognized, and presumably any captive individuals were labeled chimpanzees. Whereas Köhler abandoned research with chimpanzees after leaving Tenerife, Yerkes devoted his entire career to studying apes and promoting the field of primatology. In addition to research on tool use, problem solving, and insight (experiments that bore great similarity to Köhler's), Yerkes also examined memory, perception, communication, and virtually any other relevant aspect of ape behavior.

As well as reporting the results of his controlled studies on ape intelligence, Yerkes also summarized the known history of human interaction with great apes, which reaches back as far as the fourth and fifth centuries B.C. His writings include informed speculation about the emotions, temperament, and creative imagination of the great apes and even practical topics related to husbandry and captive management.

While trained as a psychologist, Yerkes is remembered as a devoted primatologist. His legacy includes one of the world's largest primate centers, in Atlanta, Georgia, as well as dozens of students and colleagues who became the next generation of researchers to dedicate their careers to a deeper understanding of the great apes.

WHAT IS THE AVERAGE IQ FOR AN APE?

The intelligence quotient (IQ) is a reliable means of predicting a person's academic success, and his or her score is usually consistent over a lifetime. However, even though IQ provides a meaningful scholastic rating, it doesn't reveal anything about a person's character, happiness, or social abilities. While useful for assessing academic performance, it simply doesn't measure how successful a person may be in life. The tests that are used to determine IQ reflect the educational system and society in which they are used and don't always transfer very well across cultures. These same limitations make it impossible to fairly apply tests written for humans to the nonhuman great apes. As a result, there is no standardized method for assessing the IQ of any species except *Homo sapiens*.

The general cognitive abilities of the different species of great ape have been studied to varying degrees, although the total number of individuals represented in these studies is actually very small. In addition, the overwhelming majority of the literature is based on work conducted with chimpanzees (*Pan troglodytes*), making this research the basis of most of the generalizations about great ape intelligence. There is no information to suggest that any one species of great ape is "smarter" than the other species, and all appear to have the same types of cognitive capacities. However, it is

The majority of cognitive research with great apes has focused on chimpanzees. Generalizations about great ape intelligence must be made cautiously, since differences may exist among species.

abundantly clear that a wide range of variation exists among all of the individual apes that have been studied. These differences in performance are likely to be much more significant than any of the disparities that may be confirmed to exist among species.

Given the limitations that exist for measuring great ape intelligence, there is no way to make an overall, direct comparison with human mental ability. The average nonhuman great ape cannot be assigned a "mental age" that corresponds to a human standard. A more productive way to compare the intelligence of humans and the other great apes is by examining specific cognitive skills and then looking for any similarities and differences that may exist. A basic mental ability, such as learning to recognize your own reflection in a mirror, is found in humans and other apes but not in any other type of primate. In this case, strong cognitive similarities exist. More complex features, such as those that involve logic and abstract reasoning, are clearly present for great apes but are more challenging to assess accurately among the nonhuman species. Researchers have only begun to understand the mental processes behind many of the sophisticated social behaviors exhibited by great apes, such as cooperative hunting, observational learning, and deception. In general, ape intelligence is still largely unexplored and is currently a vibrant area for research and investigation.

On the basis of the information that currently exists, it is fair to say that for certain mental skills, nonhuman apes and humans have very similar abilities. For other skills, nonhuman apes may share a specific capacity, but the level of ability demonstrated by humans is remarkably sophisticated in comparison. In some instances, when tested on equivalent tasks, nonhuman apes have exceeded the performance of their human counterparts. While cognitive similarities are evident, differences are obvious as well. However, it appears that the mental differences between humans and nonhuman great apes are primarily in degree, not in kind.

DO PRIMATES HAVE BIG BRAINS?

Brains are the control center for every function of the body. When it comes to brain size, there is a common misperception that bigger is better. While larger muscles provide more strength, a bigger brain doesn't necessarily imply greater cognitive ability. The absolute size of the brain, usually measured by weight or volume, may simply be related to overall body dimensions. Elephants have larger brains than humans, but they also have bigger hearts, livers, kidneys, and so on. Elephants aren't smarter than humans, even though their brains are much larger in direct comparison. The same logic applies to humans and the other great apes. While humans have a brain that weighs about 1.3 kilograms, and the average nonhuman great ape brain is about 0.45 kilograms, absolute size doesn't explain the difference in mental skill.

The size of the brain compared with the body, or relative brain size, is a more useful measure that allows for some generalizations. Species that have a high relative brain size usually live in complex social groups, have a varied diet, large home ranges, and a long period of dependence as they slowly mature. As a group, primates have the largest relative brain size of all mammals. Within the order, the haplorrhines (tarsiers, monkeys, and apes) have relatively larger brains than the strepsirrhines (lemurs, lorises, pottos, and galagos). Within the haplorrhines, fruit-eating species have relatively larger brains than leaf-eating species. This difference is most likely due to the increased cognitive demands that are necessary for remembering the location and timing of fruiting trees within a home range compared with the ubiquitous nature of leafy vegetation. Even though the brain is one organ, it is composed of many distinct parts. Each of its features can be affected by natural selection, leading to an emphasis on or reduction in their associated functions. In the case of frugivores, the neural architecture that supports the variety of mental challenges associated with navigating through the forest in search of food may have led to the notable differences between these species and those that subsist primarily on leaves.

A correlation also exists within the order between the size of the brain and longevity. This is especially pronounced for humans and the other great apes, the primate species that have the greatest longevity and the largest relative brain sizes. Species that live longer may experience more fluctuation in environmental conditions during their lifetimes. Adapting to these changes may require more behavioral flexibility, supported by brains that have a greater computational capacity. The offspring of these species also have a comparatively longer period of dependence. Large brains are expensive in terms of energy, complexity, and the amount of time that is required for their development. While the brain may grow to its full size during the juvenile stage, it is not structurally or functionally mature until sometime after sexual maturity is reached. This combination of factors associated with offspring development creates a responsibility for their caregivers that may last well into adulthood. By necessity, the parents must also have long life spans in order to provide properly for their young.

Aside from these broad characteristics related to primate natural history, actual and relative brain sizes reveal little else and don't provide any explanation for the range of mental abilities demonstrated by primates. In addition to large relative brain size overall, primates also have a cerebral cortex that is larger than expected. This part of the brain contains association areas that integrate information received from other structures in the brain and are involved in higher-order sensory and motor actions. One of these areas, the frontal cortex, is responsible for making plans that consider the consequences associated with the potential options. Traditionally, this area has been considered largely responsible for the differences that set the human mind apart from the other species of primate. Recent evidence has demonstrated that the size of this area does not significantly differ between humans and the other great apes, although it is smaller in the gibbons and monkeys. The most current behavioral and neurobiological evidence demonstrates that the brains of humans and other great apes have only quantitative, not qualitative, distinctions.

Rather than a single, anatomical explanation, it is much more likely that a combination of factors is responsible for the relative complexity of the human brain compared with those of the other primates. One clear factor is the surface area of the brain, which is affected by the density of the folds that are present on the cerebral cortex. Primitive brains are actually quite smooth, and more advanced brains are very convoluted. These gyri (crests) and sulci (grooves) increase the number of possible connections among nerve cells, which support more complex behaviors. The ultimate measure of brains will be derived from a thorough understanding of their internal architecture, which can vary considerably among species. Current techniques, such as electrical mapping, magnetic resonance imaging (MRI), and positron emission topography (PET), have the potential to define very specific areas of the brain and reveal their corresponding function. Research in this area of neurobiology has become an integral element in the body of information used by researchers who study the behavior and cognition of primates.

DO NONHUMAN PRIMATES USE TOOLS?

Tool use is one of the topics that is studied most often in attempting to understand behavior and cognition. Until the 1960s, it was assumed that "Man" was the only *natural* tool user. That distinction ended when Jane Goodall observed the chimpanzees of Gombe using small twigs to extract termites from their mound. Since that time, the study of primate tool use has flourished.

A variety of definitions for what constitutes tool use have been proposed, ranging from the very general to the very specific. The most widely accepted definition involves several factors. Tool use must be goal directed and involve the manipulation of an unattached object in the environment. The user establishes the proper and effective orientation of the tool and holds or carries the tool during (or just prior to) use. The tool cannot be a part of the user's body, but it can be something that is alive. Consider the following scenario: A spider monkey spots ripe fruit dangling from the tip of a branch, but unfortunately it is out of her reach. She grabs the branch with her tail and bends it toward herself until the fruit is only inches away. She plucks off the tastiest pieces with her hand and releases her tail-hold on the branch. Although a clever way to solve the problem, there is no tool use involved, since only body parts were utilized but nothing unattached. In the same scenario, tool use would be evident if the monkey had broken a branch from the tree and used that, instead of her tail, as a reaching tool to pull the fruit closer.

Behaviors that meet all of the requirements for genuine tool use are widespread throughout the primate order. While tool use has not been documented for prosimians, both New World and Old World monkeys are known to use tools occasionally. Most commonly, monkeys use tools in three general contexts. The first is for extending their reach. Many types of monkey, such as macaques (*Macaca* spp.) and baboons (*Papio* spp.), have been documented to use a variety of objects to reach and rake food or other objects toward themselves. The second category is related to exerting increased force. Capuchin monkeys (*Cebus* spp.), for example, regularly use stones or other hard objects to break open nuts or other tough foods. Colobus monkeys (*Colobus* spp.) will use digging sticks to widen the entrance areas of insect nests. By far, the last and most regular category of tool use exhibited by monkeys involves aiming or throwing objects to augment an agonistic display. Many genera of both New World and Old World monkeys throw a variety of objects such as sticks, branches, stones, sand, and gravel at humans and other intrud-

Chimpanzees commonly make and use tools for a wide variety of purposes. Here, an adult female is extracting termites from their mound with a tool she has made.

ers, but only a portion appear to aim the tool directly at the target. Other, less commonly seen, tool-using behaviors by monkeys include wiping with leaves to clean a piece of food or a wound or probing crevices with a tool in search of food.

The greatest diversity of nonhuman tool use is demonstrated by the great apes. However, a real dichotomy in tool-using behavior exists between the wild and captivity. In the wild, chimpanzees (*Pan troglodytes*) reign supreme as the master tool users. They have been documented to engage in tool-using behaviors that include "fishing" and "dipping" for insect prey, brandishing and hitting with clubs, using sponges, levers, hammers, anvils, pestles, protecting their feet while walking on thorny patches, and using cushions to sit on wet ground. All of these behaviors involved only materials from the forest environment, and a complete list of the specific types of tool use demonstrated by wild chimpanzees would simply be too long to include in this section.

Orangutans (*Pongo* spp.) use tools far less frequently in the wild. They are known to protect themselves from rain or strong sun by using detached leaves or branches, to throw objects at intruders, to scratch themselves with sticks, to use leaves for wiping or holding spiny foods, and to use small branches to swat at insects. Very recently, they have been observed to use probing tools to extract seeds from tough-skinned fruits. Orangutans also use tools to extract insects, larvae, and honey from nest holes in trees. Surprisingly, bonobos (*Pan paniscus*) and gorillas (*Gorilla* spp.) show an almost complete absence of tool use in the wild.

In captivity, both bonobos and gorillas are remarkably different from their wild counterparts, showing a wide range of tool-using skills, such as in reaching, probing, hammering, and wiping, in using objects as containers, and so on. In experimental settings, bonobos have demonstrated the ability to learn to produce stone flakes that can be used for cutting. In comparison, the range of tool-using ability of chimpanzees in captivity is similar to that demonstrated in the wild. Captive orangutans are quite distinct, having proven themselves to be the most skilled and inventive tool users of all. In addition to the tool-using skills demonstrated by the other species of nonhuman great ape, orangutans have been known to perform such activities as hanging a hammock, sweeping floors and pathways, and using a piece of wire to pick and successfully open a padlock.

Captive settings are undoubtedly influenced by the chance to interact with humans and observe human behavior, but this provides only the opportunity to learn about tools and their potential uses, not the mental capacities to perform the actions. Since chimpanzees, and orangutans to a lesser extent, are the only reliable tool users in the wild, what accounts for the clear differences in performance for orangutans, gorillas, and bonobos in captivity? The most likely explanation is that only wild chimpanzees live in environments where tool use is necessary for daily activities, such as in the acquisition or processing of food items. The other species of nonhuman great ape live in habitats where their physical abilities alone are sufficient for satisfying their most common needs. In captivity, where basic needs are supplied by humans, these apes may initiate tool use as a response to an environment that lacks consistent mental challenges or as a means of investigating situations that are not present for them in the wild. Informal tool-using activities may also be presented and encouraged by human caretakers. Experimental settings that are specifically designed to document the cognitive capacities associated with tool use have contributed substantially to what is understood about great ape abilities. It appears that the mental abilities necessary for complex forms of tool use are shared by all of the great apes and best expressed on the basis of necessity, opportunity, and circumstance.

DO NONHUMAN PRIMATES MAKE TOOLS?

Tool making is distinct from tool using. Making a tool involves the modification of an object by the user (or another group member) so that it serves more effectively as a tool. Defining certain behaviors as "tool use" may be problematic and subject to debate, but "tool making" is much more direct and easier to classify.

Neither tool using nor tool making is found in the prosimians. In the wild, both New World and Old World monkeys use tools. The majority of these are unmodified objects found in their environment, such as stones thrown at intruders or leaves that are used to clean foods. Some simple forms of modification have been observed, such as detaching a branch that is then used as a reaching tool. Less commonly, an object may be made into a tool by taking off specific parts. Capuchin monkeys (*Cebus* spp.) remove the bark from small sticks, making them more effective for probing. The results from experiments in captivity are consistent with observations from the wild. Monkeys that have been tested have shown comprehension of the direct relationship between a tool and a task, such as using a stick to obtain an out-of-reach object. Capuchin monkeys are particularly industrious in their use of tools and have been the subjects of choice for investigating tool use and manufacture. Although success and comprehension vary among individuals, the ability to modify objects in order to make them more effective as tools is widespread. In one study that required a thin, straight tool for removing food from a tube, the monkeys demonstrated two specific types of modification. In the first, they separated one stick from a large bundle of sticks that were tied together and used it successfully to remove the reward. This behavior provided evidence that the monkeys understood the bundle was too large to be used and needed to be altered. In the second example, side branches were removed from a stick so that it could fit properly into the entrance of the tube. The monkeys in this study demonstrated the ability to make an effective tool from an unsuitable object found in their environment.

Gorillas (*Gorilla* spp.) and bonobos (*Pan paniscus*) that live in the wild demonstrate a very narrow range of tool use. The only consistent example is the use of vegetation such as small trees that are thrown or dragged to augment behavioral displays. These tools are obtained by detaching them from the ground, a very simple form of modification. By comparison, orangutans (*Pongo* spp.) use tools more frequently and in a wider array of contexts. Some of their tools, such as large leaves that are used as protection from rain, are simply removed from branches and held like an umbrella. Others are carefully manufactured for a specific purpose. Some populations of orangutans use tools to extract insects, larvae, and honey from tree holes as well as to remove nutritious seeds from the inside of *Neesia* spp., a football-sized fruit. Insect-extraction tools are used for hammering and poking to gain access into the nests and as probes to remove the insects, larvae, and honey. *Neesia* fruits have a tough, spiny covering that splits open as they ripen and seeds that are protected by hairs that sting and irritate the skin. A tool is used to brush away the hairs and move the seeds toward the opening, where they can be removed using fingers or lips. Orangutans make tools that are specifically suited to these very different tasks. The seed-extraction tools are thin and short compared with the insect-extraction tools, which are long and wide. Once a seed-extraction tool is made, the orangutans carry it with them as they travel to different *Neesia* trees. Unlike the *Neesia* fruits, tree holes that contain insects can vary in both their depth and the diameter of the entrance. Orangutans appear to modify both the length and diameter of their insect-extraction tools on the basis of the size of the nest being visited. In addition to making tools that meet the requirements for a specific task, they are also able to vary the design of these tools in order to maximize their usefulness. Orangutans, therefore, have a "tool kit."

Wild chimpanzees (*Pan troglodytes*) are habitual tool users in every population that has been studied. Modified and unmodified objects are used as tools, and these great apes have the most elaborate tool kit known for any primate except humans. Chimpanzees manufacture a remarkable range of tools for such uses as sponging, hammering on an anvil, pestle pounding, puncturing, insect extraction, and dipping honey. The creation of these tools involves a variety of techniques that vary in their complexity. Simpler tools, such as sponges used to absorb drinking water, can be made by chewing a wad of leaves into the correct consistency. Others, like the wands used for termite "fishing," are more intricate. Termite fishing requires a tool that is smooth and long enough to probe inside a termite mound. It must also be flexible enough to bend around curves but still have enough strength not to break off inside the nest. Chimpanzees begin by choosing the appropriate material to work with, such as thin twigs or grassy stems. These are shortened to the right length and stripped of any side branches. The selection of raw materials and manufacturing stages of the process may occur as the apes travel toward a termite mound, clearly demonstrating anticipation of a planned tool-using activity as well as a stored mental image of a template for tool construction. During the process of fishing, the chimpanzees continue to refine the tool as the tip frays or bends.

While complex in terms of manufacture, the relationship between the fishing tool and the task is very direct. Chimpanzees are capable of tool use and manufacture that involve a more sophisticated level of comprehension related to the tool and its use. Nut cracking provides the best set of examples. Some wild chimpanzee populations commonly use stone hammers and anvils to open nuts that have exceedingly hard shells. Each chimpanzee chooses its stones carefully, and some even show a preference for the same set of stones over time. Anvil stones must have a flat, level surface on which the nut is placed. The hammer stone may be smaller and does not need to be as flat. In this type of tool use, an additional level of complexity is apparent. The ape understands the relationship between the hammer, anvil, and nut. On occasion, nut-cracking chimpanzees have shown a deeper level

Chimpanzees understand the relationship between sets of objects that can be used as tools, such as a hammer and anvil for opening hard-shelled nuts. In the wild, chimpanzees demonstrate the most varied, complex, and sophisticated forms of tool use and manufacture found in the great apes. (Photograph by Tetsuro Matsuzawa)

of understanding between the tools and the task. If a level anvil is not available, chimpanzees have used a third stone as a wedge to prop up the anvil and create a useful cracking surface. In this scenario, the ape understands the relationship between the anvil, wedge, hammer, and nut. The use of a tool to improve another tool is called metatool use. Presently, this is the most complicated form of tool use and manufacture demonstrated by any wild great ape.

Tool use and manufacture occurs in all of the major groups of primates with the exception of the prosimians. However, there are varying degrees of sophistication that have a taxonomic correlation. The majority of the tool use exhibited by monkeys involves unmodified objects that are used only once. In both wild and captive contexts, monkeys show the ability to perform only simple modifications to improve an object used as a tool. Captive and wild great apes exhibit much more diversity, flexibility, and innovation in tool use and manufacture by comparison. Great apes comprehend layered relationships between tools and their manufacture, leading to sophisticated abilities such as those seen in examples of metatool use. The differences between humans and the other great apes with regard to tool use and manufacture appear to pertain to the complexity and number of relationships that can be understood. As in all other areas of cognition that have been investigated, humans and nonhuman great apes appear to vary only in the degree of what can be understood rather than in the kinds of concepts that can be grasped.

HOW DO PRIMATES COMMUNICATE?

All primates communicate with each other in a number of different ways that vary depending on circumstance and intention. Scent, touch, hearing, and vision all function as pathways for sending and receiving information. The contents of the exchange may reveal emotional states, social plans, the presence of a predator, or a good foraging site, among multiple other possibilities. The sender of a message may be attempting to communicate with only one receiver or with an entire forest of individuals.

The loudest and most dramatic types of communication produced by primates are vocal calls that are broadcast throughout the forest. The aptly named howler monkeys (*Alouatta* spp.) as well as gibbons (*Hylobates* spp.), colobus monkeys (*Colobus* spp.), and orangutans (*Pongo* spp.) are particularly good examples. Each of these arboreal species is able to produce vocalizations that are nearly deafening at close range and can travel for great distances throughout the forest. These calls may last for many minutes and are usually performed on a daily or near-daily basis. The purpose of these calls is understood to function primarily as an advertisement of home range, which is intended to repel rivals and competitors. In orangutans, only adult males produce this type of vocalization, which is known as their long-call or great-call. In addition to announcing their location to rival males, it also acts as a means for attracting sexually receptive females. The calls, or "songs," made by the monogamous gibbons reinforce the bond between a pair in addition to defining their territory. The type of sound involved with these broadcast vocalizations travels effectively in the forest canopy, which is an environment where visual signals can't be easily seen. Nonarboreal species of primate don't utilize these types of calls but have other means for announcing their location. Chimpanzees (*Pan troglodytes*), which are much more terrestrial by comparison, use their hands and feet to drum on the buttresses of large trees. This behavior, performed mostly by males, produces a sound that can carry for long distances across the ground.

In addition to broadcast calls, primates send and receive information directly to other group members, such as alarm calls that are used to alert nearby individuals to the presence of danger. These vocalizations also draw attention to the callers, which may place them at higher risk. Therefore, they are generally used in situations that may benefit kin or members of the group. A primate that is traveling alone may not produce an alarm call if a predator is nearby, choosing to move silently away or conceal itself instead. Unlike alarm calls, contact calls occur almost constantly among group members in many species. Japanese macaques (*Macaca fuscata*) use a soft coo vocalization as a way for each member to stay in close contact with other individuals and to prevent themselves from becoming lost or separated. These are produced as an exchange among individuals, such that there is a short gap between the call and the response. The receivers are able to al-

This black howler monkey's vocalization will travel far into the forest, serving as an advertisement to potential mates and a possible warning for rivals.

ter the features of their reply and match the coo of the caller to which they are responding. In this way, they are able to indicate that their vocalization is a direct reply to one particular group member.

Other types of close-range communication that involve only a sender and a receiver can be more variable and are not limited to only one method of exchange. An ovulating macaque (*Macaca* spp.) exudes pheromones that communicate her attractiveness to any males that are nearby. If a female prefers one particular male, she has the ability to make her interest clear through gaze and body posture. Although her pheromones and appearance advertise her reproductive state, she can communicate her mate choice by approaching a particular male and presenting her swollen rear end to him (see *Why Do Some Primates Have Swollen Rears?*). This is most often accompanied by a direct glance directed toward him from over her shoulder, as added emphasis.

The face in general and the eyes in particular are a very important means of communication for primates. Stares communicate a strong message when they are aimed directly into the eyes of the receiver. Males and females both use an unflinching gaze as a way to intimidate or effectively threaten subordinates. A direct stare accompanied by an open mouth signals an even stronger warning that aggres-

The interaction between these two male chimpanzees is clearly tense. The open-mouth yawn by the individual on the right indicates an expression of anxiety or an attempt to convey a threat.

sion may be imminent. Many species of primate also use an open-mouth yawn, which exposes all of the teeth, as a way to express dominance or anxiety. It is important to note that some of the facial expressions made by humans may have entirely different meanings when exhibited by other species of primate. An obvious example is the human smile, which usually accompanies happiness. Opening the lips and showing all of the teeth clenched together is a very positive signal when exchanged between humans and is almost universally recognized. This exact same expression is an indication of extreme fear for all other species of primate (see *Which Primates Have Emotions?*).

In some situations, it may be prudent to keep communication subtle and quiet. Male chimpanzees can be very competitive with each other over access to ovulating females, and advertising interest in a particular mate may only draw the attention of rivals. Therefore, males have several ways to discreetly invite females to copulate. A male may show his interest by facing a female and revealing his erect penis. If a more dominant male comes near, "advertisers" have been known to quickly close their legs or conceal their genitals under their hands. Cultural differences have been noted as well. Males in some populations have been documented to invite copulation by swishing a small branch in the direction of a desired female. In other populations, males use "leaf clipping," in which they look toward their preferred female as they rip a leaf into pieces, which fall to the ground.

Whether through gesture, gaze, or vocalization, primates demonstrate a wide-

Vervet monkeys communicate with each other about potential predators, using a variety of calls that have symbolic meanings.

ranging ability to communicate effectively with each other. Much of this can be easily observed and accurately interpreted. It is easy to recognize the familiar comforting touch between a mother and an infant or a tug that invites play between juveniles. Facial expressions, vocalizations, and body posture are typical modes of primate communication, revealing mood, social status, and sexual interest. Many of these are common throughout the order and can be observed in every forest, grassland, office, or boardroom where primates occur.

DO ALL PRIMATES HAVE THEIR OWN LANGUAGES?

By definition, language involves communication, but not all communication involves language. While all primates certainly communicate with each other, there is no evidence to suggest that any species except *Homo sapiens* naturally uses language. There is no universally agreed-upon definition of what constitutes language, but most would agree that the minimum elements are a vocabulary of abstract symbols and syntax. Symbols may be spoken, written, or gestured and usually have no direct relationship to what they represent except for what is agreed upon by the users. For example, the words *apple* and *manzana* are completely different, yet they function as the name for the same fruit in English and Spanish. Syntax is the agreed-upon way in which the symbols in a vocabulary can be used together to express meaning. The sentences "Cats eat mice" and "Mice eat cats" are made from the same symbols but have radically different meanings because of syntax.

Symbolic communication has been documented to occur in wild primates, and the most detailed studies have involved vervet monkeys (*Chlorocebus aethiops*). These small monkeys live in social groups, and they use vocal communication to warn each other about danger. Rather than using a generalized alarm call, these monkeys have vocalizations that are specific, such as those for "leopard," "snake," and "eagle." Each of these elicits a different reaction from the monkeys, demonstrating the symbolic nature of the warnings. When the call for "eagle" is heard, the monkeys quickly run into underbrush, hiding themselves from an aerial attack. The call for "leopard" sends the monkeys scrambling into the trees, where they can easily avoid the danger. Large snakes that prey on vervets rely on concealment and stealth in order to make a kill. When the call for "snake" is heard, the monkeys simply stand up on their legs and scan the surrounding area to identify the snake's position. Once seen, the snake loses its element of surprise, and the monkey is safe.

The fact that responses to these calls are learned rather than instinctive further demonstrates the calls' symbolic nature. Young individuals are less reliable in their use of the calls, but their accuracy increases over time. Adults may even ignore the alarm calls made by inexperienced individuals. This vocabulary can also change if other types of threats are evident. A call for "unfamiliar human" has been recorded, as well as one for "baboon," which occasionally prey on vervets.

The well-studied behavior and vocalizations of these monkeys clearly demonstrate their ability to use and understand a set of abstract symbols and to develop new entries in their vocabulary. While obviously a means to communicate specific information, these vocalizations lack syntax, which distinguishes this type of communication from language. Ongoing studies in the wild and in captivity continue to explore the range of communicative abilities that naturally occur for primates. New information may reveal previously unanticipated abilities, possibly expanding the number of primate species that are considered natural language users.

CAN NONHUMAN PRIMATES LEARN A LANGUAGE?

The history of ape language research is short, and the attempted research projects are few in number. The earliest studies devoted to teaching language to apes were first reported around the turn of the twentieth century. At the time, the obvious intelligence of nonhuman great apes was recognized, and it was assumed that if treated like human children, they would be able to acquire spoken language. As a result, pioneers in the field raised infant great apes in a home environment and went to great lengths in their attempts to teach them to talk. The most successful of these projects was run by Keith and Cathy Hayes, a husband and wife team. In the early 1950s, they attempted to teach Vicki, an infant female chimpanzee, a vocabulary of spoken words. Ultimately, Vicki was able to produce breathy approximations of the words *Mama, Papa, cup,* and *up,* which she used accurately. In hindsight, these initial investigations into ape language abilities may seem burdened with unrealistic expectations and an inappropriate methodology. While that may be the case, they are very valuable in that they paved the way for groundbreaking research that would come only a few years later. The anatomical and neurological limitations that prevent apes from producing human speech do not mean that they are unable to understand and use language.

The important distinction that revealed the language ability of chimpanzees, and subsequently gorillas, orangutans, and bonobos, was freedom from a vocally based system of symbols and syntax. In the mid-1960s, Allen and Beatrix Gardner, also a husband and wife team, were the first researchers to begin an ape language study using American Sign Language (ASL). The Gardners worked with an infant chimpanzee they named Washoe, who was raised much like a human child and immersed in an environment where ASL was the primary means of communication. She and her human companions shared affectionate social interactions and made signing in ASL part of their daily routine. Over a number of years, Washoe eventually learned a vocabulary of hundreds of ASL gestures. As with human language use, she comprehended many more signs than she regularly produced. Washoe became the first nonhuman great ape to demonstrate clear comprehension of a human language. Today, Washoe and the other chimpanzees that joined the project live as a family group at the Chimpanzee and Human Communications Institute at Central Washington University and regularly communicate with their human caretakers using ASL. These chimpanzees also use sign language to communicate with each other and even sign to themselves.

The accomplishments of Washoe and her human colleagues began an era of greatly increased interest in the language and communication abilities of great apes. As a result, ape language research expanded to include gorillas, orangutans, and bonobos, and both gorillas and orangutans have learned to understand and communicate with ASL. However, the overwhelming majority of language research studies have continued to be conducted with chimpanzees, as has been the case in all other areas of behavioral and cognitive research with great apes.

Different methodologies have been devised to create nonspoken ways for apes and humans to communicate. In addition to gestures, chimpanzees and orangutans have learned to use symbols that are presented in a number of different ways, such as in the form of plastic shapes that can be arranged to construct sentences or as shapes on an interactive keyboard that both the human researchers and the chimpanzees can use to communicate with each other or to make requests. Chimpanzees have learned to send each other information via computerized keyboards in order to solve a task that requires a cooperative effort.

As technology has changed, computers have streamlined the way in which written symbols can be utilized. Two orangutans named Azy and Indah are the first of their species to learn how to communicate using abstract symbols. Using a touch-sensitive computer screen, these two individuals can name foods and objects and label quantities with numbers. Both can request actions of their human partner in these studies, and Indah is particularly good at producing short sentences with a verb and a noun, such as "open bag." At the Primate Research Institute of Kyoto University, an adult female chimpanzee named Ai uses a large vocabulary of computer-generated symbols to reveal her impressive range of mental abilities, which include the comprehension of language tasks and the use of numbers. She is the only chimpanzee that is able to use a touch-screen computer to draw her symbol answers rather than choosing them from an array. Bonobos living at the Language Research Center at Georgia State University are immersed in an environment with an interesting blend of methodologies that are used to study their range of language and other cognitive abilities. Both the humans and the nonhuman apes that work in this project use a large vocabulary of written symbols. Additionally, the humans use spoken translations for the symbols simultaneously or sometimes use spoken language alone. The bonobos have demonstrated an obvious understanding of all the forms of language that are presented to them. They are able to respond accurately to the spoken words they hear and to the written symbols they see. These great apes can also point to the symbol that is the written translation for a spoken word they have heard. In this unique situation, the humans and the bonobos are able to communicate with each other using a blend of styles, which clearly confirms a sophisticated level of comprehension.

The body of information that has been generated by this relatively recent field of research proves that humans are not the only species that can use and understand language. Current results have largely silenced skeptics and critics who generalized flaws from poorly designed studies to all ape language projects. Whether presented gesturally, vocally, or through written symbols, all of the great apes have shown the clear ability to express their thoughts and respond to human companions using language. These abilities have also been a perfect platform for exploring broader cognitive skills, such as memory, perception, and individual styles of learning. As studies in this field of research grow and mature, the results will continue to iden-

Indah, an adult female orangutan, is able to use abstract symbols and syntax to communicate her thoughts. She is pictured here with R. Shumaker. (Photograph by Richard T. Nowitz)

tify those mental abilities that are uniquely human and those that are shared with a wider number of other primate species.

WHY CAN'T APES TALK?

The production of spoken language involves a complex interaction between many parts of the body, all of which are coordinated by specific regions of the brain. In order to speak, air must move from the lungs through the larynx, over the vocal folds (sometimes inaccurately called vocal cords), and then out through the mouth. Sound is generated and shaped by the movement of air over the vocal folds and is refined into words by the mouth, lips, teeth, and tongue. The distinguishing characteristics of an individual's voice are affected by many factors, including the way in which sound resonates in the nasal cavity and other spaces in the head. The most essential aspect of speech production is the coordination of the relevant anatomical gateways in order to regulate airflow so that accurate vowel and consonant sounds are produced. Humans and the other great apes have all of the same anatomical features related to speech production, but in a slightly altered formation. This relatively small difference has profound consequences related to vocal abilities.

Human and nonhuman great ape infants have their larynx in a relatively high position in their throat. This allows them to lock the top of their larynx with the back of the nasal cavity, making it possible to breathe through the nose and swallow at the same time. At this stage of development, both are able to vocalize, but the types of sounds that can be produced are very limited. Around the age of three months, the larynx in human infants begins to move into a lower position in the throat, making it impossible to breathe and swallow simultaneously. This arrangement changes the ways in which airflow is controlled, allowing for the production of the range of sounds that are used in speech. It also increases the risk of choking on food or fluids for humans, but the danger is low compared with the benefits associated with talking. The larynx is completely descended by the end of adolescence, and by this point, the vocal tract rests at a nearly 90-degree angle behind the tongue as it enters the throat.

As nonhuman great apes mature, the position of their larynx does not descend. As a result, their range of vocalizations remains limited, and they are unable to produce consonants or complex vowels. All nonhuman great apes have a range of normal vocalizations that they use to communicate, but these involve basic vowel sounds. Some apes have produced approximations of consonants with the requisite movement of the lips, but rather than involving air through the larynx, these sounds involve air from the mouth.

Perhaps surprisingly, the most likely explanation for the difference in the arrangements of the vocal anatomy of humans and nonhuman great apes is related to the way we walk. Bipedalism requires balancing the head in an upright position on top of the spine rather than in a slightly jutted forward position, as in the quadrupedal great apes. This upright position reorients the vocal apparatus as well as increasing the mobility of the tongue. These factors suggest that bipedalism and an increase in vocal communication were linked for the ancestors of modern humans.

Most important, the anatomical features responsible for speech production are independent of the cognitive abilities that allow for language comprehension. While nonhuman great apes cannot produce spoken language, they are perfectly capable of understanding both spoken language that they hear and symbols and gestures that they see, and they can communicate in turn with those symbols and gestures. Presumably, human ancestors had the same abilities prior to becoming fully bipedal, suggesting that gestural language most likely preceded the development of spoken language.

CAN MONKEYS AND APES COUNT?

Counting is a very specific way to express information about quantities. Unlike relative judgments such as comparing "more" and "less," counting is absolute. It relies on a series of symbols, such as 8, to label specific amounts. These cardinal tags, or numbers, are used to identify specifically how many items exist in a particular array. Unlike "more versus less" decisions, counting can occur without making any comparisons to other arrays. The ability to mentally represent and manipulate concepts related to quantity has been studied in a surprisingly small number of primate species.

Two common squirrel monkeys (*Saimiri sciureus*) have learned to recognize the arabic numerals 0, 1, 3, 5, 7, and 9. After about 4,000 trials, they were able to learn how to reliably select the largest number in arrays ranging from two to four choices, earning a food reward when correct. Pairs of numerals were also shown to these monkeys. After about 800 additional trials, they showed a significant tendency to choose the pair producing the largest summed total.

Rhesus monkeys (*Macaca mulatta*) have also demonstrated an understanding of arabic numerals. They have been able to learn the values associated with the numerals 0 through 9 and can reliably choose the larger number when presented with a pair of numerals, such as 4 and 7.

Rhesus monkeys have also convincingly demonstrated the ability to order quantities from smallest to largest. Using pictures that contained differing numbers of shapes (four circles or three squares, for example), the monkeys learned how to sequence the quantities of one through four. After about 2,000 of these training trials, the monkeys were given new sets of quantities ranging from five through nine. They immediately understood how to arrange these arrays from smallest to largest.

Much research has been devoted to the numerical abilities of chimpanzees. At Georgia State University, two chimpanzees have learned to understand the values associated with the numerals 1 through 8. When they are given two pairs of numerals, such as 1 and 3, and 2 and 4, they can accurately identify the larger of the two totals. To do this, they first must mentally combine the two amounts, determine which is larger, and then make their selection.

At the Primate Research Institute of Kyoto University, a female chimpanzee named Ai has learned to understand and use the numerals 0 through 9. She can count arrays of items, such as pencils, keys, blocks, or dots on a computer screen. Using a touch-screen computer, Ai can look at a randomly arranged set of numerals, such as 1 through 6, and rearrange them in ascending order. She is just as accurate at the task if some of the numerals in the set are intentionally left out of the display. She is currently able to arrange any combination of the numerals 0 through 9 in the correct order.

Chimpanzees at the Ohio State University can count sets of objects that range in size from zero through nine items. They can also use fractions. When presented with a whole fruit, which is then sliced in front of them, they can correctly identify whether a portion of the fruit is one-quarter, one-half, or three-quarters of the original amount.

Sheba, one of the adult females in the project, is particularly skilled in the use of numbers. She can look at an array of objects and then reliably point to the numeral that correctly labels how many objects are there. Once Sheba had mastered the number values one through six, she was tested on her ability to perform addition. To do this, Sheba was presented with a set of items that were spread around a room rather than located together. The plan was to show Sheba the scattered items and then ask her to choose an answer from the group of number cards that were available. After being shown the items, Sheba reached out and selected the correct sum on her own. With absolutely no training in addition, she had spontaneously mastered the concept on the basis of her understanding of the individual number values. Even more impressive is the fact that when the items were replaced with written numerals, Sheba was equally accurate.

Currently, the majority of the studies devoted to nonhuman primates' numerical abilities have focused on squirrel monkeys, rhesus monkeys, and chimpanzees. As intriguing and impressive as the results have been, it is important to note that only 3 of the approximately 350 different species of primate have been studied. Investigations with a broader range of species will be essential for our fuller understanding of the numerical abilities that exist within the primate order.

DO NONHUMAN PRIMATES MAKE UP NEW WAYS TO SOLVE PROBLEMS?

Primates in the wild show wide diversity in their behaviors, abilities, and adaptations. A significant difficulty in understanding how often new behaviors are created and added to an individual's or a group's repertoire is directly related to how long individuals have been observed and studied. The only way to document accurately the emergence of an innovative behavior, or series of behaviors, is to know that it was absent from a group prior to the first time it was recorded by an observer. Consider the following hypothetical illustration: An anthropologist who has been studying humans for eleven months is amazed to see a very bizarre series of events. A number of young juveniles emerge from their homes after dark (a rare event) wearing masks and carrying bags. As they approach other homes, they vocalize loudly and are given preferred bits of food by the occupants. These novel behaviors, which appear to be a new food-gathering technique, are successfully repeated many times during the night. Strangely, over the next few months, this innovative technique is never seen again. If this anthropologist continues to observe humans for a number of years, it will become obvious that Halloween is an annual tradition rather than a new invention, but the origins will remain mysterious. Of course, scientists studying primate behavior in the wild are not confronted with such exaggerated scenarios. However, these researchers do rely on the collective knowledge of a population in order to make judgments about the novelty of behaviors.

Innovations may occur as a result of a variety of different circumstances, and most examples are related to foraging activity. Common behaviors may simply be redirected toward a new situation and develop into a useful means for solving a problem. Tufted capuchin monkeys (*Cebus apella*) routinely pick up and pound any items in their environment. They also demonstrate the ability to use hard objects to crack open nuts. This tool-using innovation is likely to have arisen from their normal pounding activity. In this case, the solution to the problem required the monkeys to restrict their pounding to a target, which yielded a profitable result. New behaviors may also result from a change in conditions. During a drought, one population of vervet monkeys (*Chlorocebus aethiops*) learned to soak dried *Acacia* seedpods in the exudate from the same species of tree. This novel behavior allowed the monkeys to obtain more moisture than they could have gathered with their hands alone and also softened the seedpod before it was eaten. This specific group had been observed and studied for over ten years prior to this event, making a very convincing case for the occurrence of genuine innovation.

New behaviors that accompany long-term changes in environmental conditions are more difficult to document, since occurrences of these may be sporadic and separated by many years. The relatively recent spread of oil palms (*Elaeis* spp.) throughout the range of chimpanzees (*Pan troglodytes*) has provided a unique opportunity to document the behaviors that have emerged in response to the presence of this abundant, nut-producing tree. Currently, the best (and perhaps only) example of innovative tool use that documents both emergence and dissemination within a population is "pestle pounding," performed by a group of chimpanzees in Bossou, Guinea. The apes in this population have been studied regularly, but not continuously, since 1976. This innovative tool-using behavior was first observed during 1989–1990. Chimpanzees in this population have been seen to climb the oil palm trees and stand in the crown, separating the mature leaves with their hands and feet in order to gain access to the young shoots. Once exposed, these shoots are forcefully pulled out, leaving a vertical hole. The base of the shoots are eaten, and the remaining petiole, which may be more than 3 meters long, is then used as a pestle and repeatedly pounded into the hole in the crown, producing a wet, fibrous product that is eaten by the chimpanzees. Pounding and extraction may continue until the apes can reach into the trunk up to their shoulder. This behavior may be a creative way to obtain a juicy food during the low fruiting time of the year or a way to exploit a food source in areas with a poor diversity of choices. Since it was first observed, this behavior has begun to spread throughout the population and is now practiced habitually by many members of the group.

Behavioral innovations by monkeys and apes appear infrequently and are important opportunities for researchers. These uncommon events are a means of understanding the range of behavioral flexibility and cognitive skill that exist for species that are confronted with new situations and challenges. Additionally, new behaviors may remain with only one individual or potentially disseminate to others. The rate at which innovations emerge and possibly spread throughout a group provides a window into the existing social networks as well as a potential means for assessing the degree to which individuals and species are capable of learning through observation and social interaction.

HOW DO PRIMATES LEARN FROM EACH OTHER?

Different species of primate have shown the ability to learn by observing the behavior of other individuals. "Social learning" occurs in different forms that vary in their degree of sophistication. In order of least to most complex, these forms are called social facilitation, stimulus enhancement, emulation, and imitation. The most basic form, social facilitation, describes situations where one member of a group performs a "contagious" behavior, which others then replicate. If one monkey sees a predator and begins to run, the rest of the group usually runs as well. In this case, the first monkey's behavior may simply have alerted others to the presence of danger, which facilitated their response.

Stimulus enhancement takes place when one individual's attention to a specific object increases others' attention to the same type of object. When adult chimpanzees are using hammer and anvil stones as tools to crack open nuts, youngsters routinely handle stones simultaneously. They have no idea how to use them to open nuts, but the behavior of the adults draws their attention to the stones that they will eventually learn to use as tools.

As young chimpanzees mature, they may begin to emulate adults by orienting stones in the correct hammer and anvil position and even gathering nuts without knowing how to crack them open. This illustrates their clear understanding that a connection exists among the anvil, hammer, and nuts, but the goal of nut cracking is still unclear. This set of behaviors indicates that they are emulating the adults without a firm idea of how to complete the task.

Imitation is the most complex form of social learning. It is apparent when a completely naive observer is able to perform a new behavior accurately, without any trial-and-error learning, after seeing a demonstrator perform the same behavior. Imitation implies that the observer understands all facets of a novel behavior and how these facets relate to a specific goal. Additionally, imitation may also indicate that

Social learning is one way that primates acquire information and develop new skills. Here, a young chimpanzee watches "fishing" behavior by an adult. Expertise with this form of tool use requires years of practice.

the observer has the ability to visualize a situation literally from another individual's point of view. This sophisticated ability is not commonly demonstrated by nonhuman primates. The less complicated forms of social learning are usually sufficient as explanations for the occurrence of new behaviors.

The most well-studied case of social learning in primates involves Japanese macaques (*Macaca fuscata*) living on Koshima Island. These monkeys were first observed by Japanese primatologists in 1948 and became the subjects of a field study that has lasted for more than 50 years. In the early years of the project, the researchers provided the monkeys with sweet potatoes. Provisioning had the practical benefit of bringing the monkeys into open areas where their social behaviors could be more easily observed.

In 1953, a new behavior was invented by a female macaque, who was subsequently named Imo. This one-and-a-half-year-old juvenile washed her sweet potato in a nearby freshwater brook before she ate it. The innovation became a standard practice for her, and others within the group slowly began to wash their potatoes as well. By 1958, almost 80 percent of the juveniles in the group (ages two to seven years) engaged in sweet potato washing, while only about 20 percent of the adults did so. By 1962, almost 75 percent of the group (including all adults and juveniles) showed the behavior.

During this period, juveniles from one to two-and-a-half years of age were the most likely to learn the washing behavior. The least likely were any of the oldest members of the group. Interestingly, males over four years old were slow at learning to wash their potatoes, while females of the same age were not. These trends in the dissemination of the behavior throughout the group provide clues to how it was learned. All of the youngest juveniles and females of any age had close social contact with other individuals at all times, while juvenile males around four years of age and older normally spent more time on the periphery of their group. The most basic elements necessary for social learning to occur are a competent demonstrator and an interested observer, and this clear difference in exposure is the likely explanation for the disparity in the documented learning rates. In fact, Imo's playmates (which included some young kin) were the very first to learn how to wash their sweet potatoes. Imo's mother and other kin were the next most likely monkeys to have acquired the behavior. Initially, this innovation spread from offspring to mother and from younger sibling to older sibling. Once sweet potato washing became habitual for many individuals in the group during 1958–1959, almost all infants were exposed to this behavior from birth, and the route of transmission became mother to infant. After that point, no differences in the demonstration of the behavior were seen between males and females.

When monkeys wash their sweet potatoes, small bits fall back into the water as they eat. Infants and other individuals that are nearby may pick up these pieces and consume them, promoting acquisition of the behavior. Since this washing behavior spread through the group at a relatively slow rate, imitation was unlikely the means by which it was learned. It is much more likely that this innovation spread through a combination of social facilitation and stimulus enhancement. Although these forms of social learning are basic, they can exert a powerful influence on the acquisition of new skills, which is strongly demonstrated by the effect that Imo's technique had on the entire Koshima group.

Once a behavior becomes part of a group's normal repertoire, it can also evolve on the basis of individual variation and preference, something else that was evident with the sweet potato washing. When the innovation began in 1953–1954, the potatoes were washed only in freshwater brooks. A few years later, in 1957–1958, the monkeys on this island began to avoid the brooks and use seawater instead. By 1961, both fresh and saltwater were utilized. However, freshwater was used only if

An adult Japanese macaque on Koshima Island dips a sweet potato into the sea. (Photograph by Benjamin Beck)

it was especially convenient, or it was used by subordinate monkeys trying to avoid more dominant monkeys that were near the sea. Overall, saltwater was preferred for sweet potato washing. The researchers studying this behavior proposed two possible reasons. One was that freshwater was limited on the island, and the other was that the monkeys preferred to eat their potatoes with a salty flavor. Currently, the monkeys on Koshima Island are still given sweet potatoes on an occasional basis. Unlike those in 1953, the potatoes available today are washed before they are sold. However, Imo's influence continues, because the descendants of the monkeys first studied decades ago still rush to dip their clean potatoes into the sea. Their behavior confirms what the researchers suspected: the Koshima macaques simply prefer seasoned potatoes over plain.

WHICH PRIMATES HAVE CULTURE?

Traditionally, some academic disciplines have considered "culture" to be exclusively human territory, narrowly defining it as the pattern of behaviors exhibited by *Homo sapiens* as mediated by the use of speech and language. Describing culture in this way may be useful as a framework for discussing the range of human behaviors, but it also unnecessarily removes all other species from consideration. Other scientists prefer a more general definition of culture that is not based on a particular species but focuses instead on the ways in which information is exchanged among individuals and how the behavior of groups changes over time.

Most of the researchers who study primate behavior consider culture to have a number of basic elements. Most important, culture is based on the social transmission of behaviors from one individual to another, with the possibility that these behaviors may be modified by other individuals in the future. Behaviors that are learned through trial and error or that are passed from parent to offspring through genes most likely have a noncultural explanation. The transmission of information through social learning can occur with varying degrees of sophistication, the most complex form being imitation of a completely novel behavior (see the definition of imitation in *How Do Primates Learn from Each Other?*). Direct teaching provided by a demonstrator to an observer is particularly beneficial (but not essential) for promoting culture. Both imitation and direct teaching are very common among humans. These abilities may be absent in all other primates except for the other great apes, among which they have been documented to occur only rarely.

The sweet-potato-washing Japanese macaques (*Macaca fuscata*) of Koshima Island were the first nonhuman primates in which the origins of culture were studied, and documentation of their unique behaviors set the stage for investigations of other species. This relatively new field of study has focused on the great apes, which have provided compelling evidence for the presence of cultural variation among different populations living in the same habitats. Currently, fieldwork devoted to the topic has centered primarily on orangutans (*Pongo* spp.) and chimpanzees (*Pan troglodytes*).

Geographically distinct populations of orangutans have shown clear cultural differences related to tool use. The different study sites for the populations of orangutans that have been observed are ecologically similar, containing the same foods and raw materials. Orangutans in both settings eat the calorie-rich seeds from the *Neesia* fruit, but only one population uses tools in the process (see *Do Nonhuman Primates Make Tools?*). *Neesia* fruit is covered with a tough hide, and the seeds are protected with irritating fibers that sting the skin. As the fruit matures, narrow splits open along the sides, exposing the seeds. The tool-using orangutans make a short, strong utensil that is used to remove the fibers, detach the seeds, and then maneuver them toward the opening, where they can be obtained with a finger, the tool, or even dropped right into the mouth. This population is able to remove many more seeds than the non–tool users and has a genuine advantage in exploiting *Neesia* as a food source. This same group also uses probing tools to extract insects and honey from tree holes in their forest. In the case of the tool-using orangutans, it is likely that these innovations began with creative or insightful individuals, and

Chimpanzee tool-using behaviors differ among populations. Termite fishing, pictured here, is customary in some wild populations but has apparently not been invented in others. This variation in tool-making and tool-using behaviors provides evidence of cultural variation in wild chimpanzees.

the spread of the behaviors throughout the local population resulted from social learning. These particular tool-using behaviors may simply not have been invented in the populations where they appear to be absent. Studies in the wild are able to document these important differences among populations. However, identifying the specific means by which innovations are disseminated among individuals is problematic in the wild and much harder to observe and record.

As in all other areas of cognition that have been studied, a much richer body of information exists for chimpanzees than for any of the other nonhuman species of great ape. Field studies from many different sites across Africa have confirmed clear cultural variations among chimpanzee populations. While too numerous to list, these behaviors span all aspects of chimpanzee daily life, from food collection, preparation, and consumption to the ways in which individuals interact and socialize. Among many other examples, populations show significant variation in how they use tools to crack open nuts, fish for termites, or dip for ants. These behaviors may be very common in some groups and completely absent in others, unrelated to local ecological conditions. A range of culturally specific social behaviors also exists and demonstrates that different populations have developed distinct ways of communicating agitation, friendship, and sexual interest in each other. One of the best studied of these is a behavior called hand clasp. In this local custom, two chimpanzees hold each other's hand over their heads while they use their free hands to groom each other simultaneously. Like many other behaviors, hand clasping is habitual for many chimpanzees and completely unknown to others. On the basis of collective knowledge gained from decades of research in the wild, it is clear that chimpanzees exhibit the widest range of variation and complexity in cultural behaviors currently known for any species of primate other than humans.

Questions about culture are permanently intertwined with other aspects of cognition, such as insight, creativity, and, most important, social learning. Given that these capacities are more highly developed in great apes than they are in monkeys and prosimians, it is not surprising that greater levels of cultural variation should also be present for great apes. This is not to suggest that aspects of culture are completely absent in other species of primate, only that the range found in great apes is likely to be broader by comparison. In the same way, human and nonhuman great ape cultures should not be considered as distinct entities, but rather as exhibiting gradations in complexity. There is no doubt that humans are more cognitively complex than the other great apes and that human culture is remarkable by comparison for its breadth and sophistication. The variation that is present for the nonhuman great apes cannot be compared directly but can be viewed as a window through which the behavioral and cognitive origins that have allowed human culture to flourish can be more completely understood.

ARE HUMANS THE ONLY DECEPTIVE PRIMATE?

Deception is common for many types of animals, and primates are no exception. A behavior that is performed in order to cause confusion or misinterpretation for another individual and in turn benefit the actor can be considered deceptive. These acts generally involve some form of concealment, withholding, or distraction. The simplest forms of deception are generally programmed reactions to specific situations or events, such as hiding from or misdirecting a predator. Other forms of deception indicate the presence of more complex mental skills. These behaviors have greater flexibility and can be changed depending on the circumstance or what has been learned from prior events. They may also suggest that the deceiver has some awareness of what another individual believes. The most sophisticated acts of deception are those that are based on an understanding of another's goals or intentions and are specifically constructed in order to alter what the other individual knows or believes. This level of calculation relies on the cognitive ability to comprehend a situation from another individual's mental or visual perspective, with the understanding that that perspective can be manipulated with false information. Unlike the simpler forms of deception that are widespread among primates, tactical deception on this scale is suspected to occur only among the great apes. Humans, of course, are masterful at this type of behavior and engage in it with such regularity that it is commonplace.

The simpler, reflexive forms of deception found in primates are easy to identify, observe, and elicit. Observations in the wild and studies in captivity can verify the presence of these behaviors. Sophisticated acts of deception are much more difficult to study and have yet to be conclusively documented in either the wild or captivity. Given the nature of these acts, a level of uncertainty always exists in their interpretation. They are also rare, since frequent attempts at deception are likely to be identified by other group members, leading to negative social consequences for the deceiver. As a result, descriptions of deception in nonhuman primates (both in the wild and in captivity) are usually based on single events that may have alternative explanations. However, some examples that have been witnessed and recorded by experts in primate behavior strongly suggest intentional deception. Most have been observed when they occurred spontaneously in unstructured situations, although examples from controlled settings exist as well, and all have involved chimpanzees (*Pan troglodytes*).

Acts of deception have been reported from an experimental situation involving young chimpanzees who were individually tested on a task. Each was shown two out-of-reach containers, one of which contained a preferred food and another that was empty. After the chimpanzee saw the location of the containers, a naive experimenter entered the room, and if the chimpanzee pointed to the baited container, she was given the contents. If the ape pointed to the empty container, she received nothing. The apes readily learned to indicate the location of the reward. However, an important twist was introduced in which the naive experimenter could be "friendly" or "unfriendly." The unfriendly person, who was obvious by his or her consistently worn bandana, hat, and sunglasses, would keep the food rather than share it with the ape after it had identified the proper container. The friendly person never kept the food. Over time, some of the young chimpanzee subjects began to change their responses on the basis of which person entered the room. While all eagerly showed the friendly experimenter the location of the food, they frequently pointed to the empty container for the unfriendly visitor, who left empty-handed.

Young chimpanzees in a free-ranging situation have also demonstrated behaviors that appear to be deceptive. In a study conducted in a 0.4-hectare field, experimenters wanted to learn how a single individual could affect the movements and behavior of the entire group. To do this, they selected one female, Belle, who accompanied them as they placed a pile of preferred food at a randomly selected location in the field. She was not allowed to eat the food and was then returned to her group. At that point, the entire group was allowed outside, and as Belle ran to the food, the others followed. Once they reached the baited site, all of the apes ate some portion of the reward. This routine was disrupted when a dominant individual, Rock, began to bully Belle and take the entire pile of food for himself.

The experimenters report that Belle began to devise a number of increasingly manipulative strategies in order to deceive Rock. Initially, she simply stopped uncovering the food if Rock was nearby and sat on it instead. Rock quickly caught on and pushed her off the pile, taking it all for himself. Next, Belle stopped short of the food if Rock was following. His response was to begin searching near her, which usually uncovered the reward. To counter this behavior, Belle stopped farther away and ran for the food only when Rock was distracted and looking the wrong way. Rock responded by learning to be extravigilant. On several occasions, Belle even led the group in the opposite direction, away from the food, and doubled back as they searched in vain. Once Rock was able to counter all of these strategies, the experimenters introduced a new twist. In addition to the original pile of food, they also hid a single piece of the same food nearby. Belle responded to this new situation by leading Rock directly to the single piece and then dashing for the pile while he was preoccupied. Eventually, Rock began to ignore the single piece and wait until Belle revealed the location of the larger amount. Once he could no longer be fooled, Belle appeared to abandon any novel tactics and threw a temper tantrum whenever her deceptions failed.

Adult chimpanzees have also shown behaviors that appear to be deceptive, many of them in situations that involve sex. These great apes have a promiscuous mating system, with more than one male commonly copulating with the same female when she ovulates. Although females exercise some degree of mate choice, dominant males can prevent lower-ranking males from copulating. However, this does not prevent these subordinate males from advertising their interest in a female, which is frequently done by opening their legs to display their erection. If a dominant male approaches, subordinates have been seen to quickly cover their erections with their hands, apparently attempting to conceal both their intention and their penis in order to prevent an aggressive reaction from the dominant male (see *How Do Primates Communicate?*). On the occasions when a subordinate is successful and a female accepts his invitation, both individuals quietly move out of sight and copulate in an area where they are hidden from view. Female chimpanzees normally vocalize loudly during orgasm, but during these secret matings, they have been known to remain silent. Staying out of sight during a copulation and suppressing normal vocalizations strongly suggest that these individuals understand the social repercussions of their actions. In these situations, deception may prevent aggression, but it also facilitates an opportunity for reproduction that would most likely be impossible if carried out among the watchful eyes of the group.

All of these examples of chimpanzee behavior appear to show signs of deception, although skeptics argue that this explanation involves overinterpretation. But because of the fundamental nature of deception, if it is performed well, it always has the potential to create doubt in the mind of an observer. These examples suggest that deception in primates occurs along a scale of sophistication, similar to the occurrence of other cognitive skills. Simpler forms of deception may exist for monkeys, and more complex forms for great apes. Humans, without question, are the most skilled and frequent deceivers of all the primates. There is no doubt that deception is difficult to study, both experimentally and in uncontrolled situations, and that much more research will be necessary to understand the topic more completely.

ARE ALL PRIMATES SYMPATHETIC?

Although many behaviors may appear to be sympathetic, most of us would agree that sympathy involves some level of understanding and concern for another individual's situation. Therefore, true sympathy may indicate more than nurturance, caring, or affection and can be difficult to document confidently. A sympathetic behavior demonstrates that the performer of the act has taken the mental perspective of another individual into consideration and reacts in response to that perceived situation. In some cases, it may be inferred that the actor also intends to improve the other's circumstance through his or her actions. Further, as in all other areas of cognition, sympathy most likely occurs in gradations and should not be considered as only "fully present" or "wholly absent." Careful observation and cautious interpretation are fundamental aspects of science and are always appropriate when attempting to draw conclusions about the interactions that occur among individuals. This approach is useful for collecting information that describes the range of behaviors that can be associated with specific situations, cognitive abilities, or levels of social complexity. The perceptions that motivate acts of sympathy, however, cannot be observed directly and may only be inferred.

Despite these drawbacks, numerous anecdotes suggest that humans are not the only species of primate that demonstrates expressions of sympathy. The most compelling observations come from the behavior of chimpanzees (*Pan troglodytes*) with each other and with humans. Both in the wild and in captivity, these great apes have demonstrated acts that appear to meet even the strictest definitions of sympathy.

Healthy individuals have been known to change their normal behavior in order to accommodate sick or injured group members. Reports that involve wild chimpanzees from Gombe, Tanzania, clearly illustrate the point. A young male named Freud had injured his ankle. He was unable to travel quickly and needed regular stops for rest. Both his mother and brother adjusted their speed, frequently waiting with him until he could continue. As Freud slowly healed, his brother spent time sitting with him, grooming him often, and looking at his injured ankle. In a more dramatic example, an old female that was sick and stricken with polio was barely able to travel and find food. She weakly followed her two daughters as they found a fruiting tree, but she lacked the strength to climb and pick the ripe fruits that dangled above. One of her daughters descended from the tree with one fruit in her hand and another held in her mouth. She approached her mother, placed one of the fruits beside her, and sat down nearby. Both mother and daughter then ate their food together.

In captivity, chimpanzees have demonstrated sympathy toward humans that are well known to them. Some of the most detailed and reliable examples come from the early twentieth century, when it was considered appropriate to raise infant apes in human homes. The sympathetic behavior of Toto, a juvenile chimpanzee kept by C. Kearton, is best described in the original words of the author, as reported in 1925:

> I have said . . . that the sympathy between us was not all on my side, and Toto at times showed a real sense of understanding when I in my turn needed help and affection. Never did he show this so appealingly as soon after this Christmas, when I was stricken down with a severe attack of fever. . . . He would not leave me. All day he would sit beside me, watching with a care that seemed almost maternal, and anything that I wanted he would bring me. . . . In the afternoon, he would lie down on the bed beside me, put his arm out as if to protect me, and go fast asleep. . . . It may be that some who read this book will say that friendship between an ape and a man is absurd, and that Toto, being "only an animal," cannot really have the feelings that I attribute to him. Some people may say that. They would not say it if they had felt his tenderness and seen his care as I felt and saw it at that time. He was entirely lovable. (quoted in Yerkes and Yerkes 1929, 298)

WHICH PRIMATES HAVE EMOTIONS?

Primates exhibit a range of emotions that are shared with many other types of animals. Behaviors that indicate excitement, aggression, distress, or fear, among others, are obvious and easy to interpret reliably. In addition to these basic emotional states, the behavior of some species of primate suggests that they also have feelings that are more commonly attributed only to humans. The presence of more complex emotions should not be surprising, since there is an unmistakable biological continuity among prosimians, monkeys, nonhuman apes, and humans. The physical similarities, as well as taxonomically correlated levels of cognitive sophistication, are obvious. There is no compelling reason to suggest that emotion has somehow been exempted from the same evolutionary origins and processes. Therefore, the most significant overlap in emotional similarity is likely to occur between humans and the other great apes.

Although emotions can be difficult to study in any species, observations from both the wild and captivity consistently suggest that great apes experience a wide range of feelings that can include joy, sadness, rage, and boredom. In particular, the emotional reactions of young humans and nonhuman great apes have been closely observed and compared. At very early ages, both form strong emotional bonds with their primary caregiver. Normally, this special relationship is most pronounced between a mother and her offspring, although other close kin may be included as well. During their interactions with trusted individuals, young great apes express familiar emotions. They exhibit pleasure and contentment in response to nurturing acts and smile when they are happy. Any instances of discomfort or upset are best soothed with close body contact, which sometimes includes suckling for security rather than for hunger. Normally, infants and juveniles become distressed when separated from their primary caregiver and express their displeasure by crying. Newborn humans initially cry without tears, although this changes early, and tears begin to accompany most instances of unhappiness. Except for one report to the contrary among young mountain gorillas (subspecies *Gorilla beringei beringei*), great apes are not known to produce tears when they cry. As youngsters begin to mature and experience some restrictions on their behavior, temper tantrums are a common reaction. For instance, young apes that are being weaned by their mother sometimes throw intense fits that may include screaming, throwing themselves onto the ground, and even hitting and biting if they are prevented from nursing.

The period that a young human or nonhuman great ape is dependent on its mother lasts far beyond weaning, unlike among most other types of mammals. This physical, emotional, and psychological attachment typically lasts for years and has a profound developmental influence. A well-documented example that illustrates the importance of the mother-offspring bond involves two chimpanzees, an old female named Flo and her son Flint. Flo had raised a number of offspring and was an exceptionally good mother. As an older juvenile, Flint was still extremely dependent on Flo, even nursing at an age when most chimpanzees are fully weaned and

All primates deserve the highest standards of care in captivity and conservation in the wild. However, the wide range of emotions displayed by the great apes places them in a distinct category and informs us of the weighty responsibility we have for their preservation and protection.

capable of successfully foraging for food. This situation was most likely influenced by Flo's advanced age and declining health, which prevented her from disciplining Flint and encouraging his independence. When Flint was eight-and-a-half years old, early adolescence for a male chimpanzee, Flo succumbed to her poor health and died. At this age, Flint was still remarkably attached to Flo and became deeply depressed. He seemed bewildered without her, and his movements became slow and labored. Flint also stopped eating normally, and his health quickly deteriorated. The grief that Flint experienced apparently resulted in severe psychological and physical stress. Within three-and-a-half weeks of his mother's death, he died as well.

As the example of Flo and Flint shows, the behavior of nonhuman great apes and humans may have strong similarities in situations that involve extremes of emotion. Grief and depression may be accompanied by lethargy and a loss of appetite; rage can be characterized by uncontrollable anger and acts of aggression. However, these types of situations are fairly uncommon, and the majority of emotions are far less dramatic and are usually expressed more subtly and with greater variation among species. Laughter and smiles are good examples. All of the great apes have vocal laughter that can be heard when they anticipate or engage in almost any playful interaction, such as chasing, wrestling, or tickling. While more commonly heard from youngsters, laughter is heard from adults as well, particularly in their play with infants and juveniles. Similar circumstances may elicit laughter from both humans and nonhuman great apes, but humans produce a sound that is much louder and more forceful. Laughter from great apes is softer by comparison and has a much breathier quality.

Smiles also usually accompany play behavior for nonhuman great apes but are a bit more subtle than the same human expression. Great apes generally show only their bottom teeth when they smile, keeping the top teeth covered with their upper lip. They may also close their eyes tightly as they smile if they are particularly happy or playful. Among nonhuman great apes, an expression in which all of the teeth are exposed, as in a human smile, generally indicates fear. In this instance, the broader the expression, the greater the fear. The common photos of "smiling" young apes that are used on greeting cards, calendars, and so on, actually depict individuals that are demonstrating an expression of fear.

The look that accompanies contentment is also frequently misinterpreted. Great apes generally have a very relaxed face when they are at ease, showing little or no obvious emotion. At times, their bottom lip may even droop forward in a pout. Humans observing this face commonly mistake it for sadness or dejection, when it simply shows an individual in a tranquil mood.

Emotions are difficult to quantify, and interpretation will always have some level of subjectivity on the part of the observer. At a minimum, nonhuman great apes share many of the emotions that are present for humans. They have both positive and negative feelings, but the limits of these are not well understood. However, the level of understanding that we do have of great apes today is plainly inconsistent with the standards we have used to judge them in the past. On the basis of what is known from the wild and captivity, we can be confident that every great ape is a thinking, feeling individual, capable of experiencing happiness as well as grief. This simple fact has serious ethical implications regarding the treatment of great apes by humans, and informs us of the unique responsibility that we have for their care and treatment in captivity, as well as their conservation and protection in the wild.

4

PRIMATE CONSERVATION

HOW MANY SPECIES OF PRIMATE ARE THREATENED OR ENDANGERED IN THE WILD?

A number of conservation organizations regularly compile data that report on the status of primate populations in the wild. One of the most authoritative of these is the Red List, published by the International Union for Conservation of Nature and Natural Resources (IUCN). The information presented in the Red List is based on careful analysis of factors such as the rate of decline of a species' population in the wild and the percentage of appropriate habitat that has been eliminated. These data are used to categorize primate species as critically endangered, endangered, or vulnerable on the basis of the severity of the threat they face.

While these listings have great value, each conservation association may use a different taxonomic organization or slightly dissimilar criteria to assess the status of wild populations. As a result, the information from each group's report may differ from the others, potentially leading to some confusion. The variation that exists among the information from different conservation organizations should not distract from the obvious conclusion that the overwhelming majority of all primate species in the wild are at serious risk. Development in tropical and subtropical areas of the world is proceeding rapidly, with a corresponding decrease in numbers of primates as well as reduction in usable habitat.

In rare situations, some species, such as rhesus monkeys (*Macaca mulatta*), Japanese macaques (*M. fuscata*), and sacred langurs (*Semnopithecus* spp.), are able to coexist with humans in developed or urban areas and maintain viable populations. While certainly not free from conflict or competition with people, these primates

Orangutan (*Pongo* sp.)

151

Orangutans are on the brink of extinction. Unless dramatic changes are made in Sumatra and Borneo, the wild population is predicted to disappear by 2020.

have benefited from cultural and religious tolerance. In addition, these monkeys demonstrate obvious behavioral flexibility and generalist foraging patterns that allow them to adapt and survive in these situations.

By contrast, the current situation for the orangutans of Borneo (*Pongo pygmaeus*) and Sumatra (*P. abelii*) provides a dramatic example to illustrate a spectacular conservation failure. These Asian great apes are the largest arboreal species on the planet, relying on the forests of Indonesia and Malaysia for their survival in the wild. Around 1900, the total population of orangutans in Borneo and Sumatra was estimated to be in excess of 300,000 individuals. Primarily as a result of deforestation, habitat degradation, poaching, and conflict with humans, the total population in 1997 was estimated to be approximately 12,000 orangutans in Sumatra, and 15,000 in Borneo, only 9 percent of the total population that existed in 1900. As of 2002, that number had dropped even further, with only 3,000 individuals thought to exist in Sumatra and far fewer than the 1997 estimate of 15,000 in Borneo. Conservationists and ecologists that study orangutans predict that the Sumatran population will essentially be extinct by 2010, perhaps sooner. The Bornean population is expected to follow by 2020 at the latest. Despite the grim statistics, dedicated Indonesians, Malaysians, and people from around the world have not relented in their efforts to conserve these magnificent species, two of our closest primate relatives.

WHAT ARE THE MAIN THREATS TO PRIMATES IN THE WILD?

Wild primates are likely to be threatened by any situation that involves competition or conflict with humans. The most common and persistent factor that adversely affects populations of wild primates is loss of usable habitat. Areas that are capable of supporting wild primates may be logged or mined for their commercial value, which degrades the forest. Political and economic instability increase the potential that timber and mineral acquisition will be managed for short-term profit rather than with a long-term strategy that considers the value of conservation. In most cases, these activities create such dramatic ecological alterations that primates, along with most other animals, are completely displaced. If an area of forest is clear-cut, it has little chance of regeneration and will most likely be permanently lost as natural habitat.

Countries that have existing populations of wild primates usually also have increasing human populations, and many support themselves through subsistence-based agriculture. Local people use wood from the forest for fuel, and they clear patches of natural habitat to build homes and establish garden plots and small farms to provide food and income for their families. While these individual agricultural

The main threat to primate populations is loss of habitat. Commercial logging operations provide revenue for governments, industry, and local people while permanently removing large areas of primary forest.

Families may clear small areas of forest in order to grow food to support themselves. Increasing human populations require larger amounts of land to meet their needs, placing them in direct competition with native wildlife.

holdings are smaller in scale than commercial activities are, the combined effect of an ever-increasing human presence creates a major impact. Land conversion, whether commercial or private, generally provides some form of benefit or revenue for humans while eradicating habitat for nonhuman primates.

In addition to land conversion, primates are also affected by human hunting. In the past, hunting them was limited to subsistence levels by local people and was somewhat restricted by the inaccessibility of the forest. "Bushmeat" was intended for the hunters' consumption, not for financial gain. In most situations, from a conservation point of view, hunting of this sort poses no real threat to the long-term viability of primate populations. However, that scenario has changed drastically in the very recent past, particularly in many countries in Africa. Logging companies have built roads far into the forest, and their vehicles regularly travel between these remote areas and urban centers. Logging camps may employ hundreds of people, and bushmeat from areas surrounding the camp is regularly used to feed workers. In addition, timber trucks transport the meat into towns, where it is sold for a large profit to restaurants and private individuals. While virtually any type of animal may be taken as bushmeat, the situation is particularly acute for the African great apes (chimpanzees, gorillas, and bonobos), which are being killed and eaten in record numbers. While the populations of all of the great apes are plummeting dramatically (especially the Asian orangutans), the market for gorilla, chimpanzee, and bonobo meat has evolved into a lucrative commercial enterprise. Those who are participating in the bushmeat trade are usually in great need themselves and may view any animals from the forest as either pests or direct threats that can be used as

Newly built roads allow travel far into previously unaffected areas of forest. While habitat remains, the commercial hunting that ensues has a devastating effect on primate populations.

A variety of animals, such as this cuscus, are hunted as "bushmeat." The effect on great ape populations is particularly worrisome. Unchecked, the hunting of large numbers of apes will soon lead to their extinction in the wild.

a means to generate essential income. This practice is providing revenue for people such as hunters and their families, truck drivers, restaurant owners, and private sellers in the markets. In addition to being driven by economic opportunity, the trade is related to cultural preferences that exist in some areas of Africa, where great ape meat is considered a delicacy.

The cultural norms and economic incentives that drive the demand for chimpanzee, bonobo, and gorilla meat are in direct opposition to the preservation of these species in the wild. The current demand for great ape bushmeat is completely uncontrolled, and conservationists universally agree that this hunting pressure is creating a catastrophic decline in great ape populations throughout Africa, rapidly driving these species toward extinction. Cultural habits and preferences that endanger individual species or groups of animals are obviously not limited to Africa or to the great apes. In the United States, for example, conservationists regularly warn of the effects of overfishing on wild populations such as swordfish, salmon, and Chilean sea bass. These predictions are usually met with indifference, and the general population continues to consume these foods in large quantities. Great apes that are hunted for the bushmeat trade are treated no differently from swordfish, lobster, forest antelope, or any other animal that is a highly preferred source of meat. There are no other species, however, that present the same set of ethical considerations that should be applied to chimpanzees, bonobos, gorillas, and orangutans. All of these great apes demonstrate a level of cognitive sophistication and self-awareness that distinguishes them as sharing more with humans than with any other species. They are capable of abstract reasoning, complex problem solving, and behavioral innovations that include tool use and manufacture. Great apes learn from each other and pass on cultural traditions to their offspring. They maintain social relationships that may last for a lifetime and have shown evidence of compassion in their interactions with each other and with humans. They are our closest living relatives, with mental skills that differ from ours only in degree, not in kind. Great apes cannot be reasonably compared with fish or crustaceans and should not be killed to fill a cooking pot.

The combination of forces that threaten all primates in the wild, including the destruction of forests, land conversion resulting in loss of habitat, and commercial hunting, continue to present an enormous challenge. While the conservation of species and their habitats will ultimately be decided by local people and their governments, the opportunity and responsibility to assist in that process extends far beyond the borders of each country. Conservation efforts are most effective when the global community provides support for the in situ protection of threatened species and their wild habitats.

WHAT ARE THE BEST WAYS TO SAVE PRIMATES IN THE WILD?

A number of different conservation strategies have been utilized in attempts to protect wild primates and their habitat. The most consistently effective approaches always involve a strong investment in local people, providing them with both economic and idealistic incentives that are tied to conservation initiatives. The incorporation of local people into the conservation equation realistically acknowledges that the successful protection of primates and their habitat is much more likely if these projects generate benefits such as jobs, education, and revenue for local residents. As land conversion continues at alarming rates and the potential for conflict increases, the real challenge for primate conservation is their near existence to humans living in areas that surround or include wild habitat.

A model example that incorporates these ideals is the Golden Lion Tamarin Conservation Program, which is based in a suburb of Rio de Janeiro, Brazil. This highly successful project focuses on research of both captive and wild-living tamarins, habitat restoration, and reintroduction of captive monkeys into the wild in combination with public education and awareness, economic incentives, employment, and professional training for local people. The positive conservation impact of these efforts is clear. In 1982, the wild population of golden lion tamarins was estimated at only 250 individuals. Since 1984, the Golden Lion Tamarin Conservation Program has reintroduced 153 captive-born monkeys into preserved wild habitat within the historical range of this species; these monkeys, when combined with their wild-born offspring and further descendants, translate to a net increase of about 450 individuals in the local population, which now totals about 1,200 monkeys. Most important, the day-to-day management and operation of this project is now the sole responsibility of Brazilian scientists and researchers. A large percentage of these dedicated conservationists have acquired their expertise by working in the Golden Lion Tamarin Conservation Program.

Of course, people living outside range countries also have an important role to play in the conservation of primates. The most direct course of action is to become affiliated with an organization that actively supports in situ efforts to protect primates and their habitat. While membership in the form of a financial contribution is always valuable, most groups also welcome volunteer labor to assist with any number of functions.

One of the key benefits associated with membership in a conservation organization is the ability to stay well informed of the situation in the wild in addition to receiving alerts about opportunities to support impending political legislation that may affect conservation efforts. In 2000, for example, the United States Congress passed the Great Ape Conservation Act, which provides millions of dollars in funding through 2004, funding that is available for in situ efforts that directly benefit wild populations of great apes. Without public support, this important source of funding would never have been created.

The threats facing wild populations of primates are severe, and an increasing number of challenging situations require swift and creative action to ensure the protection of some species that otherwise may not survive beyond the current generation. The dedication and experience of individuals and organizations that are devoted to finding and implementing conservation solutions have the clear potential to reverse this trend. The best of these efforts depend on local people and governments forming strong partnerships, which have proven to be a highly effective form of collaboration. Concerned people outside range countries provide crucial levels of support through their financial contributions, volunteer time, and political activity. While the current situation demands immediate attention, we have obvious success stories that provide a source of optimism and illustrate how effective global teamwork can be in protecting primates.

WHO IS WORKING TO PROTECT PRIMATES?

Many different people dedicate their lives and careers to the conservation, care, study, and protection of primates. First and foremost are the employees of governmental and privately supported wildlife conservation organizations in countries that have native populations of primates. While frequently too few in number and underfunded, these park guards, rangers, trackers, veterinarians, and educators are devoted to the preservation of primates and their habitat. Successful conservation would simply not be possible without the efforts of these highly trained professionals.

The increasing emergence of sanctuaries for primates, both those from the wild and those already in captivity, has become a new reality. Range countries that are affected by aggressive habitat loss and trade in poached animals have begun to rely on sanctuaries as a permanent home for primates that have been displaced or illegally taken from the wild. In nonrange countries, such as the United States, the number of captive primates that are in need of placement exceeds the sanctuary space that is available. These facilities have been filled with primates that have been abandoned or surrendered from the pet trade, entertainment industry, substandard zoos, and biomedical research. Sanctuary managers and their staff provide for the welfare and long-term care of individual primates that would otherwise face a very uncertain future.

A wide variety of organizations around the world are involved in efforts that directly benefit primates in the wild and in captivity. Some groups focus on individual

Park guards and trackers provide the front line of defense for wildlife conservation. These dedicated men enforce the local laws that safeguard animals and their habitat. This member of an antipoaching patrol displays snares that were found within the boundaries of a protected national park.

Organizations that are committed to the protection of primates contribute in different ways. Some are more effective with political action and public awareness, while others take a more direct approach. For example, Partners in Conservation, based at the Columbus Zoo, raises money to buy equipment and pay the salaries of park guards, such as the antipoaching patrol seen here.

species or geographical locations, while others have a more diverse approach that may include everything from monitoring captive conditions to enforcing conservation laws in the wild. Although it would be impossible to acknowledge every organization that is involved in primate conservation and welfare, the following list includes well-established groups that have a long history of successfully working to promote the interests of primates.

Jane Goodall Institute
USA Headquarters
P.O. Box 14890
Silver Spring, Maryland
20910-4890
Telephone: 301-565-0086
http://www.janegoodall.org/you/
index.html

Columbus Zoo and Aquarium
Partners in Conservation (PIC)
9990 Riverside Drive
Powell, Ohio 43065
Telephone: 614-645-3550
E-mail: picafrica@aol.com
http://www.colszoo.org

Balikpapan Orangutan Society–USA
BOS-USA
P.O. Box 2113
Aptos, California 95001
http://www.orangutan.com/index.htm

Balikpapan Orangutan Survival
  Foundation–Indonesia
P.O. Box 319, Balikpapan 76103
Samboja, Km 38
Kalimantan-Timur, Indonesia
Tel. +62 542 410365/413069/735206
E-mail: boswan@indo.net.id
http://www.redcube.nl/bos

Support for African/Asian Great Apes
  (SAGA)–Japan
Telephone: 81-568-63-0547
E-mail: saga@pri.kyoto-u.ac.jp
http://www.saga-jp.org

The American Zoo and Aquarium
  Association
8403 Colesville Road, Suite 710
Silver Spring, Maryland 20910-3314
Telephone: 301-562-0777
http://www.aza.org

The American Zoo and Aquarium
  Association
The Bushmeat Crisis Task Force
8403 Colesville Road, Suite 710
Silver Spring, Maryland 20910-3314
E-mail: info@bushmeat.org
http://www.bushmeat.org

The Dian Fossey Gorilla Fund
  International
800 Cherokee Avenue, SE
Atlanta, Georgia 30315-1440
Telephone: 800-851-0203
E-mail: 2help@gorillafund.org
http://www.gorillafund.org

International Primate Protection
  League
P.O. Box 766
Summerville, South Carolina 29484
Telephone: 843-871-2280
http://www.ippl.org

Primarily Primates, Inc.
P. O. Box 207
San Antonio, Texas 78291-0207
http://www.primarilyprimates.org

Primate Rescue Center
5087 Danville Road
Nicholasville, Kentucky 40356
kyprimate@earthlink.net
http://www.primaterescue.org

The Center for Great Apes
Box 488
Wauchula, Florida 33873
http://www.prime-apes.org

American Society of Primatologists
http://www.asp.org

Conservation International
Main Office
1919 M Street, NW, Suite 600
Washington, D.C. 20036
Telephone: 1-800-406-2306
http://www.conservation.org

The Nature Conservancy
4245 North Fairfax Drive, Suite 100
Arlington, Virginia 22203-1606
Telephone: 1-800-628-6860
E-mail: comment@tnc.org
http://nature.org

International Society of Primatologists
http://www.primate.wisc.edu/pin/
ips.html

World Wildlife Fund–U.S.
1250 24th Street, NW
Washington, D.C. 20037-1175
Telephone: 202-293-4800
http://www.worldwildlife.org

ARE CAPTIVE PRIMATES IN TROUBLE?

The ethical issues surrounding primates that live in captivity are somewhat distinct from those facing primates in the wild. Overwhelmingly, species living in captivity are self-sustaining and reproduce very successfully, resulting in stable populations that are at no risk of decline. Although we should clearly recognize the continual potential for improvement and innovation, appropriate facilities and husbandry techniques exist in captivity allowing primates to carry out a healthy range of physical and social behavior. Threats to the welfare of captive primates do not exist at the species or population level but are certainly present for individuals.

For some people, captivity for any primate is offensive and considered inappropriate. For most others, the welfare of individuals is defined by their particular circumstance and treatment. Considering the spectrum of conditions that exist in

Captive conditions for primates vary widely. Substandard facilities lack appropriate housing, husbandry practices, and enriched environments and also perpetuate negative stereotypes about primates (above left). By contrast, good zoos provide primates with interesting, challenging environments that allow them to exercise their bodies and their minds (above right). This juvenile orangutan is freely traveling between two buildings at the Smithsonian's National Zoo. Opportunities to see primates engaged in their normal behaviors inspire respect and awe in zoo visitors, which, it is hoped, will lead to increased concern for welfare in captivity and conservation in the wild.

captivity, primates living in well-regulated zoos and sanctuaries generally have the highest standards of care and the lowest incidence of inhumane treatment. Accredited zoos cooperate with one another and move primates among their institutions according to carefully managed breeding and conservation programs. No responsible zoo has collected primates from the wild for a number of decades.

By far, the largest numbers of primates in captivity exist in labs and the breeding centers that supply them with animals. In some countries, such as the United States, these facilities are legally obliged to provide basic levels of care and housing. In others, no legal protections exist, and standards can be horrendous. Whether basic standards of care are provided or not, the welfare of these individuals is secondary to the research and other forms of investigation that are being conducted.

Many primates also exist in substandard zoos, the entertainment industry, and the private pet trade. The conditions for these individuals and their care and treatment frequently range from very poor to abusive and are difficult to regulate because of the absence of effective local, state, and federal laws. In addition, these situations are generally characterized by unregulated, uncontrolled, and irresponsible breeding that is driven by profit. The numbers of primates that are involved is very difficult to document accurately. One recent estimate suggests that at least 500 chimpanzees, and probably many more, are owned by private individuals in the United States alone.

Clearly, a variety of captive situations exist for primates. It is possible to find positive, enriched environments that safeguard the welfare of individuals while providing responsible care and professional management. These ethical institutions deserve generous public and private support. Sadly, we also find situations that exploit primates for commercial gain with little regard for their quality of life. Objections voiced by animal welfare organizations that are broadly supported by the public have the potential to eliminate these substandard conditions.

HOW CAN I GET INVOLVED?

A wide variety of opportunities are available for people who are interested in becoming involved in virtually any aspect of the preservation, protection, study, and husbandry of primates. Careers devoted to the care of primates in captivity can be pursued through accredited zoos and sanctuaries. Academic fields that focus on aspects of primate behavior and conservation include biology, psychology, anthropology, and ecology. Opportunities to study primates either in the wild or in captivity generally require a graduate degree as well as a significant amount of practical knowledge and experience.

Volunteer activities and membership in primate-related organizations are other productive ways to make a positive contribution. Most zoos and sanctuaries usually have very active volunteer labor forces. Working in this capacity may include everything from assisting husbandry staff with the day-to-day care of many different types of primate, to devising and producing enrichment activities, providing public education, and assisting with ongoing research projects. Participation in conservation and welfare organizations is an invaluable way to promote the interests of both captive and wild primates. These organizations effectively tackle all issues ranging from loss of habitat in the wild to lobbying for stronger regulation and improvement of substandard primate facilities. A general search on the Internet will provide contact information for primate-related groups that involve any level of interest.

Whether in a full-time career, in volunteering a few hours a week, or in sending membership dues to a favored charity, anyone can find a beneficial and appropriate way to satisfy his or her interest in primates. Given the urgency of the situation in range countries, the next decade will be pivotal in deciding the future of many wild populations. Reliance on captivity as a safe haven is increasing, yet many situations are substandard and in need of assistance. The combination of these factors serves as a reminder that opportunities for involvement are many, and the time to act is now.

As Jane Goodall writes: "I do have hope for the future of our planet, but that hope is dependent not only on policy change in business and industry, but on huge numbers of people waking up to the danger and realizing that each of us has a role to play. . . . This is the message I want to share with people—especially children—round the world" (Lindsey 1999, 104).

APPENDIX

TAXONOMIC HIERARCHY OF PRIMATES

	Scientific Name	Common Name
	Suborder Strepsirrhini	
	Infraorder Lemuriformes	
	Family Cheirogaleidae	
Genus	*Allocebus*	hairy-eared mouse-lemur
	A. trichotis	hairy-eared mouse-lemur
Genus	*Cheirogaleus*	dwarf lemurs
	C. adipicaudatus	southern fat-tailed dwarf lemur
	C. crossleyi	furry-eared dwarf lemur
	C. major	greater dwarf lemur
	C. medius	western fat-tailed dwarf lemur
	C. minusculus	lesser gray dwarf lemur
	C. ravus	greater iron gray dwarf lemur
	C. sibreei	Sibree's dwarf lemur
Genus	*Microcebus*	mouse-lemurs
	M. murinus	gray mouse-lemur
	M. myoxinus	pygmy mouse-lemur
	M. ravelobensis	golden mouse-lemur
	M. rufus	red mouse-lemur
Genus	*Mirza*	giant mouse-lemur
	M. coquereli	giant mouse-lemur or Coquerel's mouse-lemur
Genus	*Phaner*	fork-crowned lemurs
	P. electromontis	amber mountain fork-crowned lemur
	P. furcifer	Masoala fork-crowned lemur
	P. pallescens	western fork-crowned lemur
	P. parienti	Sambirano fork-crowned lemur
	Family Lemuridae	
Genus	*Eulemur*	brown lemurs
	E. albifrons	white-fronted lemur
	E. albocollaris	white-collared lemur
	E. collaris	red-collared lemur
	E. coronatus	crowned lemur
	E. fulvus	brown lemur
	E. macaco	black lemur
	E. mongoz	mongoose lemur
	E. rubriventer	red-bellied lemur
	E. rufus	red-fronted lemur
	E. sanfordi	Sanford's lemur
Genus	*Hapalemur*	lesser gentle lemurs or bamboo lemurs
	H. alaotrensis	Alaotran gentle lemur
	H. aureus	golden gentle lemur or golden bamboo lemur

	H. griseus	gray gentle lemur or gray bamboo lemur
	H. occidentalis	Sambirano gentle lemur
Genus	*Lemur*	ring-tailed lemur
	L. catta	ring-tailed lemur
Genus	*Prolemur*	greater bamboo lemur
	P. simus	broad-nosed gentle lemur or greater bamboo lemur
Genus	*Varecia*	ruffed lemurs
	V. rubra	red ruffed lemur
	V. variegata	black-and-white ruffed lemur

Family Megaladapidae

Genus	*Lepilemur*	sportive lemurs
	L. dorsalis	black-striped sportive lemur
	L. edwardsi	Milne-Edward's sportive lemur
	L. leucopus	white-footed sportive lemur
	L. microdon	small-toothed sportive lemur
	L. mustelinus	weasel lemur
	L. ruficaudatus	red-tailed sportive lemur
	L. septentrionalis	northern sportive lemur

Family Indridae

Genus	*Indri*	indri
	I. indri	indri
Genus	*Avahi*	woolly indris or avahis
	A. laniger	eastern avahi
	A. occidentalis	western avahi
Genus	*Propithecus*	sifakas
	P. coquereli	Coquerel's sifaka
	P. deckenii	Van der Decken's sifaka
	P. diadema	diademed sifaka or simpoon
	P. edwardsi	Milne-Edward's sifaka
	P. perrieri	Perrier's sifaka
	P. tattersalli	Tattersall's sifaka
	P. verreauxi	Verreaux's sifaka

Infraorder Chiromyiformes

Family Daubentoniidae

Genus	*Daubentonia*	aye-aye
	D. madagascariensis	aye-aye

Infraorder Loriformes

Family Loridae

Genus	*Arctocebus*	angwantibos
	A. aureus	golden angwantibo
	A. calabarensis	calabar angwantibo
Genus	*Loris*	slender lorises
	L. lydekkerianus	gray slender loris
	L. tardigradus	red slender loris
Genus	*Nycticebus*	slow lorises
	N. bengalensis	Bengal slow loris
	N. coucang	Sunda slow loris
	N. pygmaeus	pygmy slow loris
Genus	*Perodicticus*	potto
	P. potto	potto

Genus	*Pseudopotto*	false potto
	P. martini	false potto
Family Galagonidae		
Genus	*Euoticus*	needle-clawed bushbabies
	E. elegantulus	southern needle-clawed bushbaby
	E. pallidus	northern needle-clawed bushbaby
Genus	*Galago*	lesser bushbabies
	G. alleni	Bioko Allen's bushbaby
	G. cameronensis	Cross River Allen's bushbaby
	G. demidoff	Prince Demidoff's bushbaby
	G. gabonensis	Gabon Allen's bushbaby
	G. gallarum	Somali bushbaby
	G. granti	Grant's bushbaby
	G. matschiei	dusky bushbaby
	G. moholi	Moholi bushbaby
	G. nyasae	Malawi bushbaby
	G. orinus	Uluguru bushbaby
	G. rondoensis	Rondo bushbaby
	G. senegalensis	Senegal bushbaby
	G. thomasi	Thomas's bushbaby
	G. udzungwensis	Uzungwa bushbaby
	G. zanzibaricus	Zanzibar bushbaby
Genus	*Otolemur*	greater galagos or thick-tailed bushbabies
	O. crassicaudatus	brown greater galago
	O. garnettii	northern greater galago
	O. monteiri	silvery greater galago
Surborder Haplorrhini		
Infraorder Tarsiiformes		
Family Tarsiidae		
Genus	*Tarsius*	tarsiers
	T. bancanus	Horsfield's tarsier or western tarsier
	T. dianae	Dian's tarsier
	T. pelengensis	Peleng tarsier
	T. pumilus	pygmy tarsier
	T. sangirensis	Sangihe tarsier
	T. spectrum	spectral tarsier
	T. syrichta	Philippine tarsier
Infraorder Simiiformes		
Section Platyrrhini		
Family Cebidae		
Genus	*Callimico*	Goeldi's marmoset
	C. goeldii	Goeldi's marmoset
Genus	*Callithrix*	marmosets
	C. argentata	silvery marmoset
	C. aurita	buffy-tufted marmoset
	C. chrysoleuca	gold-and-white marmoset
	C. emiliae	Emilia's marmoset
	C. flaviceps	buffy-headed marmoset
	C. geoffroyi	white-headed marmoset
	C. humeralifera	Santarem marmoset

	C. humilis	Roosmalen's dwarf marmoset
	C. intermedia	Hershkovit's marmoset
	C. jacchus	common marmoset
	C. kuhlii	Wied's marmoset
	C. leucippe	white marmoset
	C. marcai	Marca's marmoset
	C. mauesi	Maues marmoset
	C. melanura	black-tailed marmoset
	C. nigriceps	black-headed marmoset
	C. penicillata	black-tufted marmoset
	C. pygmaea	pygmy marmoset
Genus	Cebus	capuchin monkeys
	C. albifrons	white-fronted capuchin
	C. apella	tufted capuchin
	C. capucinus	white-headed capuchin
	C. kaapori	kaapori capuchin
	C. libidinosus	black-striped capuchin
	C. nigritus	black capuchin
	C. olivaceus	weeper capuchin
	C. xanthosternos	golden-bellied capuchin
Genus	Leontopithecus	lion tamarins
	L. caissara	Superagui lion tamarin
	L. chrysomelas	golden-headed lion tamarin
	L. chrysopygus	black lion tamarin
	L. rosalia	golden lion tamarin
Genus	Saguinus	tamarins
	S. bicolor	pied tamarin
	S. fusciollis	brown-mantled tamarin
	S. geoffroyi	Geoffroy's tamarin
	S. graellsi	Graell's tamarin
	S. imperator	emperor tamarin
	S. inustus	mottle-faced tamarin
	S. labiatus	white-lipped tamarin
	S. leucopus	white-footed tamarin
	S. martinsi	Martin's tamarin
	S. midas	red-handed tamarin
	S. melanoleucus	white-mantled tamarin
	S. mystax	mustached tamarin
	S. niger	black tamarin
	S. nigricollis	black-mantled tamarin
	S. oedipus	cottontop tamarin or Pinche tamarin
	S. pileatus	red-capped tamarin
	S. tripartitus	golden-mantled tamarin
Genus	Saimiri	squirrel monkeys
	S. boliviensis	black-capped squirrel monkey
	S. oerstedti	Central American squirrel monkey
	S. sciureus	common squirrel monkey
	S. ustus	bare-eared squirrel monkey
	S. vanzolinii	black squirrel monkey

Family Nyctipithecidae
   Genus *Aotus* — night monkeys or owl monkeys or douroucoulis
      A. azarae — Azara's night monkey
      A. hershkovitzi — Hershkovitz's night monkey
      A. lemurinus — gray-bellied night monkey
      A. miconax — Peruvian night monkey
      A. nancymaae — Nancy Ma's night monkey
      A. nigriceps — black-headed night monkey
      A. trivirgatus — three-striped night monkey
      A. vociferans — Spix's night monkey

Family Pitheciidae
   Genus *Cacajao* — uakaris
      C. calvus — bald uakari
      C. melanocephalus — black-headed uakari
   Genus *Callicebus* — titis
      C. baptista — Baptista Lake titi
      C. brunneus — brown titi
      C. cinerascens — ashy black titi
      C. cupreus — coppery titi
      C. donacophilus — white-eared titi
      C. hoffmannsi — Hoffmann's titi
      C. medemi — black-handed titi
      C. modestus — Rio Beni titi
      C. moloch — red-bellied titi
      C. oenanthe — Rio Mayo titi
      C. olallae — Olalla brother's titi
      C. ornatus — ornate titi
      C. pallescens — white-coated titi
      C. personatus — Atlantic titi
      C. torquatus — collared titi
   Genus *Chiropotes* — bearded sakis
      C. albinasus — white-nosed saki
      C. satanas — black bearded saki
   Genus *Pithecia* — saki monkeys
      P. aequatorialis — equatorial saki
      P. albicans — white-footed saki
      P. irrorata — Rio Tapajos saki
      P. monachus — monk saki
      P. pithecia — white-faced saki

Family Atelidae
   Genus *Alouatta* — howler monkeys
      A. belzebul — red-handed howler
      A. caraya — brown howler
      A. coibensis — Coiba Island howler
      A. guariba — red-and-black howler
      A. macconnelli — Guyanan red howler
      A. nigerrima — Amazon black howler
      A. palliata — mantled howler
      A. pigra — Guatemalan black howler
      A. sara — Bolivian red howler
      A. seniculus — Venezuelan red howler

Genus	*Ateles*	spider monkeys
	A. belzebuth	white-fronted spider monkey
	A. chamek	Peruvian spider monkey
	A. fusciceps	black-headed spider monkey
	A. geoffroyi	Geoffroy's spider monkey
	A. hybridus	brown spider monkey
	A. marginatus	white-cheeked spider monkey
	A. paniscus	red-faced spider monkey
Genus	*Brachyteles*	woolly spider monkeys or muriquis
	B. arachnoids	southern muriqui
	B. hypoxanthus	northern muriqui
Genus	*Lagothrix*	woolly monkeys
	L. cana	gray woolly monkey
	L. lagotricha	brown woolly monkey
	L. lugens	Colombian woolly monkey
	L. poeppigii	silvery woolly monkey
Genus	*Oreonax*	yellow-tailed woolly monkey
	O. flavicauda	yellow-tailed woolly monkey or Hendee's woolly monkey

Section Catarrhini
  Family Cercopithecidae
  Subfamily Cercopithecinae

Genus	*Allenopithecus*	swamp monkey
	A. nigroviridis	Allen's swamp monkey
Genus	*Cercocebus*	white-eyelid mangabeys
	C. agilis	agile mangabey
	C. atys	sooty mangabey
	C. chrysogaster	golden-bellied mangabey
	C. galeritus	Tana River mangabey
	C. sanjei	Sanje mangabey
	C. torquatus	collared mangabey or red-capped mangabey
Genus	*Cercopithecus*	guenons
	C. albogularis	Syke's monkey
	C. ascanius	red-tailed monkey
	C. campbelli	Campbell's mona
	C. cephus	mustached guenon
	C. denti	Denti's mona
	C. diana	diana monkey
	C. doggetti	silver monkey
	C. dryas	dryas monkey or salongo monkey
	C. erythrogaster	white-throated guenon
	C. erythrotis	red-eared guenon
	C. hamlyni	Hamlyn's monkey or owl-faced monkey
	C. kandti	golden monkey
	C. lhoesti	L'Hoest's monkey
	C. lowei	Lowe's monkey
	C. mitis	blue monkey
	C. mona	mona monkey
	C. neglectus	De Brazza's monkey
	C. nictitans	greater spot-nosed monkey

	C. petaurista	lesser spot-nosed monkey
	C. pogonias	crested mona
	C. preussi	Preuss's monkey
	C. roloway	roloway monkey
	C. sclateri	Sclater's guenon
	C. solatus	sun-tailed monkey
	C. wolfi	Wolf's mona
Genus	*Chlorocebus*	vervet monkeys
	C. aethiops	grivet
	C. cynosuros	malbrouck
	C. djamdjamensis	Bale Mountains vervet
	C. pygerythrus	vervet monkey
	C. sabaeus	green monkey
	C. tantalus	tantalus monkey
Genus	*Erythrocebus*	patas monkey
	E. patas	patas monkey
Genus	*Lophocebus*	crested mangabeys
	L. albigena	gray-cheeked mangabey
	L. aterrimus	black crested mangabey
	L. opdenboschi	Opdenbosch's mangabey
Genus	*Macaca*	macaques
	M. arctoides	stump-tailed macaque or bear macaque
	M. assamensis	Assam macaque
	M. cyclopsis	Formosan rock macaque
	M. fascicularis	crab-eating macaque or long-tailed macaque or kera
	M. fuscata	Japanese macaque
	M. hecki	Heck's macaque
	M. leonine	northern pig-tailed macaque
	M. maura	Moor macaque
	M. mulatta	rhesus monkey
	M. nemestrina	Sunda pig-tailed macaque or beruk
	M. nigra	Celebes crested macaque or black "ape"
	M. nigrescens	Gorontalo macaque
	M. ochreata	booted macaque
	M. pagensis	Mentawai macaque or bokkoi
	M. radiata	bonnet macaque
	M. silenus	lion-tailed macaque
	M. sinica	Toque macaque
	M. sylvanus	Barbary macaque
	M. thibetana	Tibetan macaque or Milne-Edward's macaque
	M. tonkeana	Tonkean macaque
Genus	*Mandrillus*	mandrills
	M. leucophaeus	drill
	M. sphinx	mandrill
Genus	*Miopithecus*	talapoins
	M. ogouensis	gabon talapoin
	M. talapoin	Angolan talapoin
Genus	*Nasalis*	proboscis monkey
	N. larvatus	proboscis monkey

Genus	*Papio*	baboons
	P. anubis	olive baboon or Anubis baboon
	P. cynocephalus	yellow baboon
	P. hamadryas	Hamadryas baboon or sacred baboon or mantled baboon
	P. papio	Guinea baboon
	P. ursinus	Chacma baboon
Genus	*Theropithecus*	gelada
	T. gelada	gelada

Subfamily Colobinae

Genus	*Colobus*	black-and-white colobus
	C. angolensis	Angola colobus
	C. guereza	mantled guereza
	C. polykomos	king colobus
	C. satanas	black colobus
	C. vellerosus	ursine colobus
Genus	*Piliocolobus*	red colobus
	P. badius	western red colobus
	P. foai	Central African red colobus
	P. gordonorum	Uzungwa red colobus
	P. kirkii	Zanzibar red colobus
	P. pennantii	Pennant's colobus
	P. preussi	Preuss's red colobus
	P. rufomitratus	Tana River red colobus
	P. tephrosceles	Ugandan red colobus
	P. tholloni	Thollon's red colobus
Genus	*Presbytis*	surilis
	P. chrysomelas	sarawak surili
	P. comata	Javan surili
	P. femoralis	banded surili
	P. frontata	white-fronted langur
	P. hosei	Hose's langur
	P. melalophos	Sumatran surili
	P. natunae	Natuna Island surili
	P. potenziani	Mentawai langur or Joja
	P. rubicunda	maroon leaf monkey
	P. siamensis	white-thighed surili
	P. thomasi	Thomas's langur
Genus	*Procolobus*	olive colobus
	P. verus	olive colobus
Genus	*Pygathrix*	doucs
	P. cinerea	gray-shanked douc
	P. nemaeus	red-shanked douc
	P. nigripes	black-shanked douc
Genus	*Rhinopithecus*	snub-nosed monkeys
	R. avunculus	Tonkin snub-nosed langur
	R. bieti	black snub-nosed monkey
	R. brelichi	gray snub-nosed monkey
	R. roxellana	golden snub-nosed monkey
Genus	*Semnopithecus*	sacred langurs or gray langurs or Indian langurs
	S. ajax	Kashmir gray langur

S. dussumieri	southern plains gray langur
S. entellus	northern plains gray langur
S. hector	Tarai gray langur
S. hypoleucos	black-footed gray langur
S. priam	tufted gray langur
S. schistaceus	Nepal gray langur
Genus *Simias*	pig-tailed langur
S. concolor	pig-tailed langur or simakobu
Genus *Trachypithecus*	lutungs
T. auratus	Javan lutung
T. barbei	Tenasserim lutung
T. cristatus	silvery leaf monkey or silvery lutung
T. delacouri	Delacour's langur
T. ebenus	Indochinese black langur
T. francoisi	Francois's langur
T. geei	Gee's golden langur
T. germaini	Indochinese lutung
T. hatinhensis	Hatinh langur
T. johnii	Nilgiri langur
T. laotum	Laotian langur
T. obscurus	dusky leaf monkey or spectacled leaf monkey
T. phayrei	Phayre's leaf monkey
T. pileatus	capped langur
T. poliocephalus	white-headed langur
T. shortridgei	Shortridge's langur
T. vetulus	purple-faced langur
Family Hylobatidae	
Genus *Hylobates*	gibbons
H. agilis	agile gibbon
H. albibarbis	Bornean white-bearded gibbon
H. concolor	concolor gibbon or black crested gibbon
H. gabriellae	red-cheeked gibbon
H. hainanus	Hainan gibbon
H. hoolock	Hoolock gibbon
H. klossii	Kloss gibbon or bilou
H. lar	Lar gibbon or white-handed gibbon
H. leucogenys	northern white-cheeked gibbon
H. moloch	silvery gibbon
H. muelleri	Muller's bornean gibbon
H. pileatus	pileated gibbon
H. siki	southern white-cheeked gibbon
H. syndactylus	siamang
Family Hominidae	
Genus *Gorilla*	gorillas
G. beringei	eastern gorilla
G. gorilla	western gorilla
Genus *Pan*	chimpanzees
P. paniscus	bonobo or pygmy chimpanzee or gracile chimpanzee
P. troglodytes	chimpanzee or common chimpanzee
Genus *Pongo*	orangutans

	P. abelii	Sumatran orangutan or mawas
	P. pygmaeus	Bornean orangutan or maias
Genus	*Homo*	humans
	H. sapiens	modern human

GLOSSARY

absolute brain size The actual size of a brain, usually expressed by a measure of weight or volume. Compare **relative brain size**.

allogrooming A form of physical contact that occurs between individuals. This behavior has several functions, which include building and maintaining alliances, expressing rank, showing affection, and cleaning. Compare **autogrooming**.

anthropocentric Referring to the process of understanding and interpreting situations from an exclusively human-oriented point of view.

anthropomorphic Referring to the inappropriate application of human characteristics, motivations, or abilities to nonhuman species.

ape Species that are members of the families Hylobatidae and Hominidae, as distinct from all other primates.

aunting Care and nurturance provided by a female, other than the mother, to any infant of her species, regardless of their kinship.

autogrooming Self-directed cleaning that is specifically related to hygiene. This behavior does not involve any other individual or any form of social contact. Compare **allogrooming**.

bilateral symmetry The condition found in many types of animals, including primates, where the right and left sides of the body are comparable in their appearance.

binomial nomenclature The system invented by Carolus Linneaus in which every type of organism is assigned a scientific name. These names derive from the taxonomy of a given organism, denoting both its genus and species, such as *Homo sapiens* for humans.

bipedalism An upright posture or form of locomotion that relies on the use of the legs without any assistance from the arms. Humans are the only species of primate that habitually use a bipedal form of locomotion.

brachiation A form of locomotion that relies on the arms without any assistance from the legs. The gibbons are the only true brachiators, although some other arboreal primates are considered semi- or modified brachiators.

catarrhine Literally, "down-nosed." Any species of primate native to areas in the Old World, such as Asia and Africa.

cheek pouch Elastic cheeks that allow relatively large amounts of food to be held in the mouth temporarily, usually during foraging; found only in Old World primates.

cognition In regard to primates, a wide variety of mental processes including, but not necessarily limited to, learning, memory, understanding, perception, and problem solving.

DNA (deoxyribonucleic acid) The sequence of nucleic acids that is commonly referred to as the genetic "blueprint" for each individual organism. DNA is one means of assessing the relative similarities that exist among species.

ecological niche The role that an organism plays, especially involving its way of life and its effect on the environment, in relation to the community of species that exist in a specific habitat.

emulation The form of social, or observational, learning in which one individual (an observer) devises its own solution in attempting to accomplish a goal that has been achieved by another individual (a demonstrator). Emulation is considered more complex than social facilitation and stimulus enhancement but less complex than imitation. Compare **imitation, social facilitation,** and **stimulus enhancement**.

epitheliochorial placentation The condition in which the mother's blood vessels remain completely separate from those of her developing offspring during pregnancy. In primates, this is the norm for prosimians. Compare **hemochorial placentation**.

evolution The process by which the characteristics of a population change over time in relation to the demands of the environment. Evolution is the foundation upon which all of biology rests and is the scientific explanation for the diversity of life on Earth.

faunivorous Referring to organisms that have a diet composed primarily of animals.

fitness In evolutionary terms, a measure that refers to the number of surviving offspring left by an individual, not a measure of physical health.

folivorous Referring to organisms that have a diet composed primarily of leaves or other forms of vegetation; frequently used interchangeably with "leaf-eating."

frugivorous Referring to organisms that have a diet composed primarily of fruits.

haplorrhine Any of the "dry-nosed" primates. The haplorrhine group includes the tarsiers, monkeys, and apes.

hemochorial placentation The condition in which the mother's blood intermingles directly with that of her developing offspring during pregnancy. In primates, this is the norm for tarsiers, monkeys, and apes. Compare **epitheliochorial placentation**.

imitation The most complex form of social, or observational, learning, in which one individual (an observer) exactly reproduces the goal-oriented behavior of another individual (a demonstrator). Also known as imitative copying.

infraorder A classification further distinguishing members of an order, but more broadly than by its families.

ischial callosities Hard, thickened areas of hairless skin that are prominent on the rear ends of many species of Old World primates. These structures function primarily as sitting pads.

micromanipulation A specialized foraging technique in which the hand is inserted into holes and crevices to search for small animal or insect prey. This behavior is most commonly associated with the tamarins.

natural selection In evolutionary terms, the process that determines which physical and behavioral characteristics are best suited to the environment, leading to their persistence in future generations.

neocortex The convoluted surface area of the brain that is associated with many of the most complex mental abilities, such as language.

neotenous Referring to physical features that are usually associated with infants and juveniles, such as a relatively large, round head and short, pudgy body.

neuroanatomy The study of the structural makeup of nerves or nervous tissue, such as the organization, connectivity, and complexity of nerve cell pathways associated with specific areas or functions of the brain.

New World monkeys All species of nonhuman primates that are native to the southernmost portion of North America (Mexico) and Central and South America. "New World monkey" is used interchangeably with "platyrrhine."

Old World monkeys All species of monkey that are native to areas outside the Americas, chiefly Africa and Asia. Prosimians and apes may be correctly labeled Old World primates, but they are not monkeys. "Old World monkey" is used interchangeably with "catarrhine."

olfactory Referring to the sense of smell.

ovulation The process during a female's sexual cycle when a ripe egg is released and becomes available for fertilization.

pectoral Situated near or related to the area of the chest.

pendactyly The presence of five digits on the hand or the foot.

platyrrhine Literally, "flat-nosed" or "broad-nosed." Any species of primate native to areas of the New World, such as Central and South America.

polyandry A mating system in which one adult female simultaneously maintains more than one male mate.

prehensile For primates, referring to tails that can function as a third hand or foot by grasping, reaching, or holding. Prehensile tails are found only in the platyrrhines (New World monkeys).

primate Any member of the order Primate. All primates have a shared set of physical features, such as forward-facing eyes, opposable digits, and relatively large brains for mammals of their size.

primatology The study of any member of the order Primate.

prosimian Literally, "pre-monkey." The more primitive species of primate, including the lemurs, lorises, pottos, and galagos. Some sources include tarsiers as well, although this is debated among taxonomists. Monkeys and apes are not members of the prosimian group.

quadrumanous Referring to the unique way in which orangutans locomote using any combination of their hands and feet interchangeably.

quadrupedal Referring to the most common form of locomotion found among primates, in which the hands and feet are used simultaneously, resulting in all four limbs supporting the body during movement.

relative brain size The proportion of the body made up by the brain. Compare **absolute brain size**.

sagittal crest A flat, bony protrusion rising from the top of the skull. This structure is most exaggerated in male gorillas and serves as an attachment point for the muscles associated with chewing.

sexual dichromatism The difference in hair or skin color that is correlated with the two sexes within a given species. In primates, the adult coloration of northern white-cheeked gibbons presents an excellent example.

sexual dimorphism The difference in body size or structure that is correlated with the two sexes within a given species. For example, adult male orangutans may be twice the size of adult females.

sexual swelling The dramatic change in size and coloration of a female primate's sexual skin that is associated with ovulation; frequently used interchangeably with "tumescence."

social facilitation The simplest form of social, or observational, learning, in which one individual (an observer) is more likely to perform a specific behavior when another individual (a demonstrator) performs the same or similar behavior.

stimulus enhancement The form of social, or observational, learning in which one individual (an observer) is more attentive to some object in the environment as a result of another individual (a demonstrator) showing an interest in the same type of object.

strepsirrhine Any of the "wet-nosed" primates. The strepsirrhine group includes, among others, the lemurs, lorises, pottos, and the aye-aye, all of which are prosimians.

tapetum lucidum cellulosum The structure found on the back of the retina that reflects light, present only in strepsirrhine primates.

taxonomy The classification system introduced by Carolus Linnaeus that is based on how closely or distantly species are related to each other.

tumescence The swelling of a female primate's sexual skin that occurs around the time of ovulation.

REFERENCES

1. PRIMATES IN GENERAL

How Are Primates Classified?

Ankel-Simons, F. 2000. *Primate anatomy*. San Diego: Academic Press.

Groves, C. 2001. *Primate taxonomy*. Washington, D.C.: Smithsonian Institution Press.

Swindler, D. R. 1998. *Introduction to the primates*. Seattle: University of Washington Press.

What Are the Different Types of Primate?

Ankel-Simons, F. 2000. *Primate anatomy*. San Diego: Academic Press.

Cartmill, M. 1974. *Daubentonia, Dactylopsila,* woodpeckers and klinorhynchy. In *Prosimian biology,* ed. R. D. Martin, G. A. Doyle, and A. C. Walker. Pittsburgh: University of Pittsburgh Press.

Cheney, D. L., and R. M. Seyfarth. 1990. *How monkeys see the world.* Chicago: University of Chicago Press.

Coimbra-Filho, A. F., and R. A. Mittermeier. 1977. Tree-gouging, exudate-eating and the "short-tusked" condition in *Callithrix* and *Cebuella.* In *The biology and conservation of the Callitrichidae,* ed. D. G. Kleiman. Washington D.C.: Smithsonian Institution Press.

Erickson, C. J. 1991. Percussive foraging in the aye-aye, *Daubentonia madagascariensis. Animal Behavior* 41: 793-801.

———. 1994. Tap-scanning and extractive foraging in aye-ayes, *Daubentonia madagascariensis. Folia Primatologica* 62: 125–35.

Fleagle, J. G. 1999. *Primate adaptation and evolution*. San Diego: Academic Press.

Groves, C. 2001. *Primate taxonomy*. Washington, D.C.: Smithsonian Institution Press.

McCune, L. 1999. Children's transition to language. In *The origins of language*, ed. B. J. King. Santa Fe: School of American Research Press.

Nowak, R. M. 1999. *Walker's primates of the world*. Baltimore: Johns Hopkins University Press.

Rowe, N. 1996. *The pictorial guide to the living primates*. New York: Pogonias Press.

Stanford, C. B. 1998. *Chimpanzee and red colobus: The ecology of predator and prey*. Cambridge, Mass.: Harvard University Press.

Swindler, D. R. 1998. *Introduction to the primates*. Seattle: University of Washington Press.

Wolfe, L. D., and E. H. Peters. 1987. History of the freeranging rhesus monkeys (*Macaca mulatta*) of Silver Springs. *Florida Scientist* 50, no. 4: 234–45.

What Makes a Primate a Primate?

Ankel-Simons, F. 2000. *Primate anatomy*. San Diego: Academic Press.

Conroy, G. C. 1990. *Primate evolution*. New York: W. W. Norton and Company.

Fleagle, J. G. 1999. *Primate adaptation and evolution*. San Diego: Academic Press.

Martin, R. D. 1995. Phylogenetic aspects of primate reproduction: The context of advance maternal care. In *Motherhood in human and nonhuman primates*, ed. C. R. Pryce, R. D. Martin, and D. Skuse. Basel: Karger.

Napier, J., revised by Russell H. Tuttle. 1993. *Hands*. Princeton, N.J.: Princeton University Press.

Nowak, R. M. 1999. *Walker's primates of the world*. Baltimore: Johns Hopkins University Press.

Swindler, D. R. 1998. *Introduction to the primates*. Seattle: University of Washington Press.

Are Monkeys and Apes the Same Thing?

Swartz, K. B., D. Sarauw, and Sian Evans. 1999. Comparative aspects of mirror self-recognition in great apes. In *The mentalities of gorillas and orangutans: Comparative perspectives*, ed. S. T. Parker, R. W. Mitchell, and H. L. Miles. Cambridge: Cambridge University Press.

Ujheli, M., B. Merker, P. Bulk, and T. Geissmann. 2000. Observations on the behavior of gibbons (*Hylobates leucogenys, H. gabrielle,* and *H. lar*) in the presence of mirrors. *Journal of Comparative Psychology* 114: 253–62.

How Do Primates Move?

Erikson, G. E. 1963. Brachiation in the New World monkeys and in anthropoid apes. *Symposia of the Zoological Society of London* 10: 135–64.

Fleagle, J. G. 1974. Dynamics of a brachiating siamang [*Hylobates (Symphalangus) syndactylus*]. *Nature* 248: 259–60.

———. 1999. *Primate adaptation and evolution*. San Diego: Academic Press.

Napier, J. R., and P. H. Napier. 1985. *The natural history of the primates*. Cambridge, Mass.: MIT Press.

Rose, M. D. 1989. New postcranial specimens of catarrhines from the Middle Miocene Chinji Formation, Pakistan: Descriptions and a discussion of proximal humeral functional morphology in anthropoids. *Journal of Human Evolution* 18: 131–62.

Swindler, D. R. 1998. *Introduction to the primates*. Seattle: University of Washington Press.

White, T. D. 1991. *Human osteology*. San Diego: Academic Press.

How Closely Are Humans Related to Other Primates?

Caccone, A., and J. R. Powell. 1989. DNA divergence among hominoids. *Evolution* 43: 925–42.

Caccone, A., R. DeSalle, and J. R. Powell. 1988. Calibration of the changing thermal stability of DNA duplexes and degree of base pair mismatch. *Journal of Molecular Evolution* 27: 212–16.

Did Humans Evolve from Apes?

Alcock, J. 2001. *Animal behavior*. Sunderland, Mass.: Sinauer Associates.

Will Chimpanzees Evolve into Humans?

de Waal, F. 1982. *Chimpanzee politics: Power and sex among apes*. New York: Harper and Row Publishers.

What Are the Smallest and Largest Primates?

Fleagle, J. G. 1999. *Primate adaptation and evolution*. San Diego: Academic Press.

Nowak, R. M. 1999. *Walker's primates of the world*. Baltimore: Johns Hopkins University Press.

Rowe, N. 1996. *The pictorial guide to the living primates*. New York: Pogonias Press.

What Do Primates Eat?

Ankel-Simons, F. 2000. *Primate anatomy*. San Diego: Academic Press.

Do All Primates Have Tails?

Ankel-Simons, F. 2000. *Primate anatomy*. San Diego: Academic Press.

Ledley, F. D. 1982. Evolution and the human tail. *New England Journal of Medicine* 306, no. 20: 1212–15.

Rowe, N. 1996. *The pictorial guide to the living primates*. New York: Pogonias Press.

White, T. D. 1991. *Human osteology*. San Diego: Academic Press.

How Long Do Primates Live?

Dittus, W. P. J. 1975. Population dynamics of the toque monkey, *Macaca sinica*. In *Socioecology and psychology of primates*, ed. R. H. Tuttle. The Hague: Mouton Publishers.

Dunbar, R. I. M. 1987. Demography and reproduction. In *Primate societies*, ed. B. B. Smuts, D. L. Cheney, R. M. Seyfarth, R. W. Wrangham, and T. T. Struhsaker. Chicago: University of Chicago Press.

Richard, A. F. 1985. *Primates in nature*. New York: W. H. Freeman and Company.

Rowe, N. 1996. *The pictorial guide to the living primates*. New York: Pogonias Press.

Which Primates Are the Fastest?

Napier, J. R., and P. H. Napier. 1985. *The natural history of the primates*. Cambridge, Mass.: MIT Press.

Nowak, R. M. 1999. *Walker's primates of the world*. Baltimore: Johns Hopkins University Press.

Rowe, N. 1996. *The pictorial guide to the living primates*. New York: Pogonias Press.

How Strong Is a Gorilla?

Willoughby, D. P. 1978. *All about gorillas*. Cranbury, N.J.: A. S. Barnes and Company.

2. PRIMATE SOCIAL BEHAVIOR

Why Do Primates Live in Different Types of Groups?

Barton, R. A. 2000. Primate brain evolution: Cognitive demands of foraging or of social life? In *On the move: How and why animals travel in groups*, ed. S. Boinski and P. A. Garber. Chicago: University of Chicago Press.

Chapman, C. A., and L. J. Chapman. 2000. Determinants of group size in primates: The importance of travel costs. In *On the move: How and why animals travel in groups*, ed. S. Boinski and P. A. Garber. Chicago: University of Chicago Press.

Dunbar, R. 1996. *Grooming, gossip, and the evolution of language*. Cambridge, Mass.: Harvard University Press.

van Schaik, C. P., and J. A. R. A. M. van Hooff. 1983. On the ultimate causes of primate social systems. *Behaviour* 85: 91–117.

Are Primates Monogamous?

Baker, A. J., J. M. Dietz, and D. G. Kleiman. 1993. Behavioural evidence for monopolization of paternity in multi-male groups of golden lion tamarins. *Animal Behaviour* 46: 1091–103.

Nowak, R. M. 1999. *Walker's primates of the world*. Baltimore: Johns Hopkins University Press.

van Schaik, C. P., and R. I. M. Dunbar. 1990. The evolution of monogamy in large primates: A new hypothesis and some crucial tests. *Behaviour* 115, nos. 1–2: 30–62.

van Schaik, C. P., and J. A. R. A. M. van Hooff. 1983. On the ultimate causes of primate social systems. *Behaviour* 85: 91–117.

Do All Primates Recognize Their Relatives, and Does This Influence Their Society?

Cheney, D. L., and R. M. Seyfarth. 1990. *How monkeys see the world.* Chicago: University of Chicago Press.

de Waal, F. B. M. 1996. Macaque social culture: Development and perpetuation of affiliative networks. *Journal of Comparative Psychology* 110, no. 2: 147–54.

Parr, L. A., and F. B. M. de Waal. (1999). Visual kin recognition in chimpanzees. *Nature* 399: 647–48.

How Does Body Size Affect the Lives of Primates?

Fleagle, J. G. 1985. Size and adaptation in primates. In *Size and scaling in primate biology,* ed. W. L. Jungers. New York: Plenum Press.

How Long Do Babies Stay with Their Mothers?

Altmann, J. 1980. *Baboon mothers and infants.* Cambridge, Mass.: Harvard University Press.

———. 1993. Costs of reproduction in baboons (*Papio cynocephalus*). In *Behavioral energetics: The cost of survival in vertebrates,* ed. W. P. Aspey and S. I. Lustick. Columbus: Ohio State University Press.

Gomendio, M. 1995. Maternal styles in Old World primates: Their adaptive significance. In *Motherhood in human and nonhuman primates,* ed. C. R. Pryce, R. D. Martin, and D. Skuse. Basel: Karger.

Mittermeier, R. A., I. Tattersall, W. R. Konstant, D. M. Meyers, and R. B. Mast. 1994. *Lemurs of Madagascar.* Washington, D.C.: Conservation International.

Nowak, R. M. 1999. *Walker's primates of the world.* Baltimore: Johns Hopkins University Press.

Why Do Some Primates Have Swollen Rears?

Brown, L., R. W. Shumaker, and J. Downhower. 1995. Do primates experience sperm competition? *American Naturalist* 146, no. 2: 302–6.

Dixson, A. F. 1998. *Primate sexuality.* Oxford: Oxford University Press.

When Do Primates Mate?

de Waal, F., and F. Lanting. 1997. *Bonobo: The forgotten ape.* Berkeley: University of California Press.

Fox, E. 2001. Homosexual behavior in wild Sumatran orangutans. *American Journal of Primatology* 55, no. 3: 177–81.

Do Males or Females Initiate Mating?

Dixson, A. F. 1998. *Primate sexuality: Comparative studies of the prosimians, monkeys, apes, and human beings.* Oxford: Oxford University Press.

Goodall, J. 1986. *The chimpanzees of Gombe: Patterns of behavior.* Cambridge: Cambridge University Press.

Keddy, A. C. 1986. Female mate choice in vervet monkeys. *American Journal of Primatology* 10: 125–43.

Sicotte, P. 2001. Female mate choice in mountain gorillas. In *Mountain gorillas*, ed. M. M. Robbins, P. Sicotte, and K. Stewart. Cambridge: Cambridge University Press.

Welker, C., H. Hohmann, and C. Schfer-Witt. 1990. Significance of kin relations and individual preferences in the social behavior of *Cebus apella*. *Folia Primatologica* 54: 166–70.

Will Primates Adopt a Baby?

Goodall, J. 2003. Personal communication.

Lyttle, J. 1997. *Gorillas in our midst*. Columbus: Ohio State University Press.

Thierry, B., and J. R. Anderson. 1986. Adoption in anthropoid primates. *International Journal of Primatology* 7: 196–216.

How Have Ideas about Primates Changed?

Carpenter, C. R. 1963. Societies of monkeys and apes. In *Primate social behavior*, ed. C. H. Southwick. Princeton, N.J.: D. Van Nostrand Company.

Emlen, J. T., Jr., and G. B. Schaller. 1963. In the home of the mountain gorilla. In *Primate social behavior*, ed. C. H. Southwick.. New Jersey: D. Van Nostrand Company.

Garner, R. L. 1892. *The speech of monkeys*. London: William Heinemann.

Yerkes, R. M., and A. W. Yerkes. 1929. *The great apes*. New Haven: Yale University Press.

Zuckerman, S. 1963. Human sociology and the sub-human primates. In *Primate social behavior*, ed. C. H. Southwick. New Jersey: D. Van Nostrand Company.

Why Do Primates Spend So Much Time Grooming?

Altmann, J. 1980. *Baboon mothers and infants*. Cambridge, Mass.: Harvard University Press.

de Waal, F. 1989. *Peacemaking among primates*. Cambridge, Mass.: Harvard University Press.

Fairbanks, L. A., and M. T. McGuire. 1984. Relationships of vervet monkeys with sons and daughters from one through three years of age. *Animal Behavior* 33: 40–50.

Fobes, J. L., and J. E. King. 1982. *Primate behavior*. New York: Academic Press.

Silk, J. B., A. Samuels, and P. Rodman. 1981. The influence of kinship, rank, and sex on affiliation and aggression between adult female and immature bonnet macaques (*Macaca radiata*). *Behaviour* 78: 111–77.

Which Primates Hunt for Meat?

Goodall, J. 1986. *The chimpanzees of Gombe patterns of behavior*. Cambridge, Mass.: Belknap Press of Harvard University Press.

Harding, R. S. O. 1975. Meat-eating and hunting in baboons. In *Socioecology and psychology of primates*, ed. R. H. Tuttle. The Hague: Mouton Publishers.

Stanford, C. B. 1998. *Chimpanzee and red colobus: The ecology of predator and prey*. Cambridge, Mass.: Harvard University Press.

Do All Primates Have Friends?

Altmann, J. 1980. *Baboon mothers and infants*. Cambridge, Mass.: Harvard University Press.

Carpenter, C. R. 1963. Societies of monkeys and apes. In *Primate social behavior*, ed. C. H. Southwick. Princeton, N.J.: D. Van Nostrand Company.

de Waal, F. 1996. *Good natured: The origins of right and wrong in humans and other animals.* Cambridge, Mass.: Harvard University Press.

Smuts, B. B. 1985. *Sex and friendship in baboons.* Cambridge, Mass.: Harvard University Press.

Washburn, S. L., and I. DeVore. 1963. The social life of baboons. In *Primate social behavior,* ed. C. H. Southwick. Princeton, N.J.: D. Van Nostrand Company.

Do Primates "Make Up" after a Fight?

Cords, M., and F. Aureli. 2000. Reconciliation and relationship qualities. In *Natural conflict resolution,* ed. F. Aureli and F. B. M. de Waal. Berkeley: University of California Press.

de Waal, F. 1989. *Peacemaking among primates.* Cambridge, Mass.: Harvard University Press.

Kummer, H. 1978. On the value of social relationships to nonhuman primates: A heuristic scheme. *Social Science Information* 17, nos. 4/5: 687–705.

van Schaik, C. P., and F. Aureli. 2000. The natural history of valuable relationships. In *Natural conflict resolution,* ed. F. Aureli and F. B. M. de Waal. Berkeley: University of California Press.

3. PRIMATE INTELLIGENCE

When Was Primate Intelligence First Studied?

Garner, R. L. 1892. *The speech of monkeys.* London: William Heinemann.

Köhler, W. 1925. *The mentality of apes.* London: Kegan Paul, Trench, Trubner and Co.

Teuber, M. L. 1994. The founding of the primate station, Tenerife, Canary Islands. *American Journal of Psychology* 107: 551–81.

Yerkes, R. M., and A. W. Yerkes. 1929. *The great apes.* New Haven: Yale University Press.

Do Primates Have Big Brains?

Allman, J., T. McLaughlin, and A. Hakeem. 1993. Brain weight and life-span in primate species. *Proceedings of the National Academy of Sciences* 90: 118–22.

Ankel-Simons, F. 2000. *Primate anatomy.* San Diego: Academic Press.

Clutton-Brock, T., and P. Harvey. 1980. Primates, brains, and ecology. *Journal of Zoology* 190: 309–23.

Dunbar, R. 1996. *Grooming, gossip, and the evolution of language.* Cambridge, Mass.: Harvard University Press.

Hakeem, A., G. R. Sandoval, M. Jones, and J. Allman. 1996. Brain and life span in primates. In *Handbook of the psychology of aging,* 4th ed., ed. J. E. Birren, and K. W. Schaie. San Diego: Academic Press.

Jerison, H. J. 1973. *Evolution of the brain and intelligence.* New York: Academic Press.

Kandel, E. R., J. H. Schwartz, and T. M. Jessell. 1995. *Essentials of neural science and behavior.* Norwalk, Conn.: Appleton and Lange.

Semendeferi, K., A. Lu, N. Schenker, and H. Damasio. 2002. Humans and great apes share a large frontal cortex. *Nature Neuroscience* 5: 272–76.

Swindler, D. R. 1998. *Introduction to the primates.* Seattle: University of Washington Press.

Do Nonhuman Primates Use Tools?

Beck, B. B. 1975. Primate tool behavior. In *Socioecology and psychology of primates,* ed. R. H. Tuttle. The Hague: Mouton Publishers.

———. 1980. *Animal tool behavior: The use and manufacture of tools by animals.* New York: Garland STPM Press.

de Waal, F., and F. Lanting. 1997. *Bonobo: The forgotten ape.* Berkeley: University of California Press.

Fox, E. A., A. F. Sitompul, and C. P. van Schaik. 1999. Intelligent tool use in wild Sumatran orangutans. In *The mentalities of gorillas and orangutans*, ed. S. T. Parker, R. W. Mitchell, and H. L. Miles. Cambridge: Cambridge University Press.

Linden, E. 1999. *The parrot's lament: And other true tales of animal intrigue, intelligence, and ingenuity.* New York: E. P. Dutton.

McGrew, W. C. 1992. *Chimpanzee material culture: Implications for human evolution.* Cambridge: Cambridge University Press.

Matsuzawa, T. 2001. Primate foundations of human intelligence: A view of tool use in nonhuman primates and fossil hominids. In *Primate origins of human cognition and behavior*, ed. T. Matsuzawa. Tokyo: Springer-Verlag.

O'Malley, R. C., and W. C. McGrew. 2000. Oral tool use by captive orangutans (*Pongo pygmaeus*). *Folia Primatologica* 71: 334–41.

Tomasello, M., and J. Call. 1997. *Primate cognition.* New York: Oxford University Press.

Whiten, A., J. Goodall, W. C. McGrew, T. Nishida, V. Reynolds, Y. Sugiyama, C. E. G. Tutin, R. W. Wrangham, and C. Boesch. 1999. Cultures in chimpanzees. *Nature* 399, no. 6737: 682–85.

Do Nonhuman Primates Make Tools?

Beck, B. B. 1975. Primate tool behavior. In *Socioecology and psychology of primates*, ed. R. H. Tuttle. The Hague: Mouton Publishers.

———. 1980. *Animal tool behavior: The use and manufacture of tools by animals.* New York: Garland STPM Press.

Byrne, R. W. 1995. *The thinking ape: Evolutionary origins of intelligence.* Oxford: Oxford University Press.

Fox, E. A., A. F. Sitompul, and C. P. van Schaik. 1999. Intelligent tool use in wild Sumatran orangutans. In *The mentalities of gorillas and orangutans*, ed. S. T. Parker, R. W. Mitchell, and H. L. Miles. Cambridge: Cambridge University Press.

Matsuzawa, T. 2001. Primate foundations of human intelligence: A view of tool use in nonhuman primates and fossil hominids. In *Primate origins of human cognition and behavior*, ed. T. Matsuzawa. Tokyo: Springer-Verlag.

Parker, S. T., and M. L. McKinney. 1999. *Origins of intelligence: The evolution of cognitive development in monkeys, apes, and humans.* Baltimore: Johns Hopkins University Press.

Visalberghi, E., and L. Limongelli. 1996. Acting and understanding: Tool use revisited through the minds of capuchin monkeys. In *Reaching into thought: The minds of the great apes*, ed. A. Russon, K. Bard, and S. T. Parker. Cambridge: Cambridge University Press.

Yamakoshi, G. 2001. Ecology of tool use in wild chimpanzees: Toward reconstruction of early Hominid evolution. In *Primate origins of human cognition and behavior*, ed. T. Matsuzawa. Tokyo: Springer-Verlag.

How Do Primates Communicate?

Altmann, S. A. 1967. The structure of primate social communication. In *Social communication among primates*, ed. S. A. Altmann. Chicago: University of Chicago Press.

de Waal, F. 1982. *Chimpanzee politics: Power and sex among apes.* New York: Harper and Row Publishers.

Goodall, J. 1986. *The chimpanzees of Gombe: Patterns of behavior.* Cambridge: Cambridge University Press.

Nadler, R. D. 1988. Sexual and reproductive behavior. In *Orang-utan biology*, ed. J. H. Schwartz. New York: Oxford University Press.

Sugiura, H. 2001. Vocal exchange of coo calls in Japanese macaques. In *Primate origins of human cognition and behavior*, ed. T. Matsuzawa. Tokyo: Springer-Verlag.

Do All Primates Have Their Own Languages?

Cheney, D. L., and R. M. Seyfarth. 1990. *How monkeys see the world.* Chicago: University of Chicago Press.

King, B. J. 1999. Primatological perspectives on language. In *The origins of language*, ed. B. J. King. Santa Fe: School of American Research Press.

Can Nonhuman Primates Learn a Language?

Fouts, R. 1983. Chimpanzee language and elephant tails: A theoretical synthesis. In *Language in primates*, ed. J. de Luce and H. T. Wilder. New York: Springer-Verlag.

Fouts, R., and S. T. Mills. 1997. *Next of kin: What chimpanzees have taught me about who we are.* New York: William Morrow and Company.

Linden, E. 1974. *Apes, men, and language.* New York: Saturday Review Press/E. P. Dutton.

Miles, H. L. 1983. Apes and language: The search for communicative competence. In *Language in primates*, ed. J. de Luce and H. T. Wilder. New York: Springer-Verlag.

Patterson, F., and E. Linden. 1981. *The education of Koko.* New York: Holt, Rinehart and Winston.

Premack, D., and A. J. Premack. 1983. *The mind of an ape.* New York: W. W. Norton and Company.

Savage-Rumbaugh, S., and R. Lewin. 1994. *Kanzi: The ape at the brink of the human mind.* New York: John Wiley and Sons.

Savage-Rumbaugh, S., S. G. Shanker, and T. J. Taylor. 1998. *Apes, language, and the human mind.* New York: Oxford University Press.

Why Can't Apes Talk?

Bradshaw, J., and L. Rogers. 1993. *The evolution of lateral asymmetries, language, tool use, and intellect.* San Diego: Academic Press.

Savage-Rumbaugh, S., and R. Lewin. 1994. *Kanzi: The ape at the brink of the human mind.* New York: John Wiley and Sons.

Can Monkeys and Apes Count?

Boysen, S. T. 1997. *Representation of quantities by apes.* Advances in the Study of Behavior 26. New York: Academic Press.

Boysen, S. T., and G. G. Berntson. 1989. Numerical competence in a chimpanzee. *Journal of Comparative Psychology* 103: 23–31.

———. 1995. Responses to quantity: Perceptual versus cognitive mechanisms in chimpanzees (*Pan troglodytes*). *Journal of Experimental Psychology: Animal Behavior Processes* 21, no. 1: 82–86.

Boysen, S. T., K. L. Mukobi, and G. G. Berntson. 1999. Overcoming response bias using symbolic representations of number by chimpanzees (*Pan troglodytes*). *Animal Learning and Behavior* 27, no. 2: 229–35.

Boysen, S. T., G. G. Berntson, M. B. Hannan, and J. T. Cacioppo. 1996. Quantity-based interference and symbolic representations in chimpanzees (*Pan troglodytes*). *Journal of Exploratory Psychology: Animal Behavior Processes* 22, no. 1: 76–86.

Brannon, E., and H. S. Terrace. 1998. Ordering of the numerosities 1 to 9 by monkeys. *Science* 282: 746–49.

Kawai, N., and T. Matsuzawa. 2000. Numerical memory span in a chimpanzee. *Nature* 403: 39–40.

Matsuzawa, T. 1985. Use of numbers by a chimpanzee. *Nature* 315: 57–59.

Olthof, A., C. M. Iden, and W. Roberts. 1997. Judgments of ordinality and summation of number symbols by squirrel monkeys (*Saimiri sciureus*). *Journal of Exploratory Psychology: Animal Behavior Processes* 23, no. 3: 325–39.

Rumbaugh, D. M., and D. A. Washburn. 1993. Counting by chimpanzees and ordinality judgments by macaques in video-formatted tasks. In *The development of numerical competence: Animal and human models*, ed. S. T. Boysen and E. J. Capaldi. Hillsdale, N.J.: Lawrence Erlbaum Associates.

Do Nonhuman Primates Make Up New Ways to Solve Problems?

Boesch, C. 1996. Three approaches for assessing chimpanzee culture. In *Reaching into thought: The minds of the great apes*, ed. A. E. Russon, K. A. Bard, and S. T. Parker. Cambridge: Cambridge University Press.

Hauser, M. D. 1988. Invention and social transmission: New data from wild vervet monkeys. In *Machiavellian intelligence: Social expertise and the evolution of intellect in monkeys, apes, and humans*, ed. R. Byrne and A. Whiten. Oxford: Clarendon Press.

McGrew, W. C. 1992. *Chimpanzee material culture: Implications for human evolution*. Cambridge: Cambridge University Press.

Visalberghi, E., and D. M. Fragaszy. 1996. Do monkeys ape? In *Reaching into thought: The minds of the great apes*, ed. A. Russon, K. Bard, and S. T. Parker. Cambridge: Cambridge University Press.

Yamakoshi, G., and Y. Sugiyama. 1995. Pestle-pounding behavior of wild chimpanzees at Bossou, Guinea: A newly observed tool-using behavior. *Primates* 36: 489–500.

How Do Primates Learn from Each Other?

Hirata, S., K. Watanabe, and M. Kawai. 2001. "Sweet-potato washing" revisited. In *Primate origins of human cognition and behavior*, ed. T. Matsuzawa. Tokyo: Springer-Verlag.

Nishida, T. 1987. Local traditions and cultural transmission. In *Primate societies*, ed. B. B. Smuts, D. L. Cheney, R. M. Seyfarth, R. W. Wrangham, and T. T. Struhsaker. Chicago: University of Chicago Press.

Which Primates Have Culture?

Boesch, C. 1996. Three approaches for assessing chimpanzee culture. In *Reaching into thought: The minds of the great apes*, ed. A. E. Russon, K. A. Bard, and S. T. Parker. Cambridge: Cambridge University Press.

Kummer, H. 1971. *Primate societies*. Chicago: Aldine.

McGrew, W. C. 1992. *Chimpanzee material culture: Implications for human evolution*. Cambridge: Cambridge University Press.

McGrew, W. C., and C. E. G. Tutin. 1978. Evidence for a social custom in wild chimpanzees? *Man* 13: 234–51.

Parker, S. T., and A. E. Russon. 1996. On the wild side of culture and cognition in the great apes. In *Reaching into thought: The minds of the great apes*, ed. A. E. Russon, K. A. Bard, and S. T. Parker. Cambridge: Cambridge University Press.

van Schaik, C. P., and C. D. Knott. 2001. Geographic variation in tool use on *Neesia* fruits in orangutans. *American Journal of Anthropology* 114: 331–42.

Whiten, A., J. Goodall, W. C. McGrew, T. Nishida, V. Reynolds, Y. Sugiyama, C. E. G. Tutin, R. W. Wrangham, and C. Boesch. 1999. Cultures in chimpanzees. *Nature* 399, no. 6737: 682–85.

Wrangham, R. W., W. C. McGrew, F. B. M. de Waal, and P. G. Heltne, eds. 1994. *Chimpanzee cultures*. Cambridge, Mass.: Harvard University Press.

Are Humans the Only Deceptive Primate?

Anderson, J. R. 1996. Chimpanzee and capuchin monkeys: Comparative cognition. In *Reaching into thought: The minds of the great apes*, ed. A. Russon, K. Bard, and S. T. Parker. Cambridge: Cambridge University Press.

Byrne, R., and A. Whiten. 1988. Tactical deception of familiar individuals in baboons. In *Machiavellian intelligence: Social expertise and the evolution of intellect in monkeys, apes, and humans*, ed. R. Byrne and A. Whiten. Oxford: Clarendon Press.

de Waal, F. 1982. *Chimpanzee politics: Power and aex among apes*. New York: Harper and Row Publishers.

———. 1986. Deception in the natural communication of chimpanzees. In *Deception: Perspective on human and nonhuman deceit*, ed. R. W. Mitchell, and N. S. Thompson. New York: State University of New York Press.

———. 1996. *Good natured: The origins of right and wrong in humans and other animals*. Cambridge, Mass.: Harvard University Press.

Menzel, E. 1974. A group of young chimpanzees in a one-acre field. In *Behaviour of nonhuman primates*, vol. 5, ed. A. M. Shrier and F. Stollnitz. New York: Academic Press.

Premack, D. 1988. "Does the chimpanzee have a theory of mind?" revisited. In *Machiavellian intelligence: Social expertise and the evolution of intellect in monkeys, apes, and humans*, ed. R. Byrne and A. Whiten. Oxford: Clarendon Press.

Premack, D., and A. J. Premack. 1983. *The mind of an ape*. New York: W. W. Norton and Company.

Are All Primates Sympathetic?

de Waal, F. 1996. *Good natured: The origins of right and wrong in humans and other animals*. Cambridge, Mass.: Harvard University Press.

Goodall, J. 1986. *The chimpanzees of Gombe: Patterns of behavior*. Cambridge: Cambridge University Press.

Kearton, C. 1925. My friend Toto: The adventures of a chimpanzee and the story of his journey from the Congo to London. As reported in R. M. Yerkes and A. W. Yerkes. 1929. *The great apes*. New Haven: Yale University Press.

Which Primates Have Emotions?

Fossey, D. 1983. *Gorillas in the mist*. Boston: Houghton Mifflin Company.

Goodall, J. 1986. *The chimpanzees of Gombe: Patterns of behavior*. Cambridge: Cambridge University Press.

———. 1990. *Through a window*. Boston: Houghton Mifflin Company.

Ladygina-Kohts, N. N. 2002. *Infant chimpanzee and human child: A classic 1935 comparative study of ape emotions and intelligence*. New York: Oxford University Press.

Yerkes, R. M., and A. W. Yerkes. 1929. *The great apes*. New Haven: Yale University Press.

4. PRIMATE CONSERVATION

How Many Species of Primate Are Threatened or Endangered in the Wild?

Rijksen, H. D., and E. Meifaard. 1999. *Our vanishing relative*. Dordrecht: Kluwer Academic Publishers.

How Can I Get Involved?

Lindsey, Jennifer. 1999. *Jane Goodall 40 years at Gombe: A tribute to four decades of wildlife research, education, and conservation*. New York: Stewart, Tabori & Chang.

TAXONOMIC INDEX

Page numbers in boldface indicate illustrations.

Acacia spp., 135
angwantibos (*Arctocebus* spp.),
  9, 64
apes, 33–36
Artiodactyla, 36
Atelidae, 16, 18–19
aye-aye (*Daubentonia
  madagascariensis*), 7, 59

baboons (*Papio* spp.) 21, 25,
  26–27, 47, **62, 106,** 118;
  Hamadryas (*P. hamadryas*), 26;
  olive baboon (*P. anubis,*) **11,
  13, 26, 27, 43, 73, 74, 87,**
  100, 102, **103;** yellow baboon
  (*P. cynocephalus*), 82
bats (order Chiroptera), 36
bonobo (*Pan paniscus*), 35, **35,**
  43, 52, 55, 56, 90, 120–121,
  129–130, 154–156
bushbaby, **5, 9,** 9–10

capuchin monkeys (*Cebus* spp.),
  15, 16, 18, 62, **63,** 91, 118,
  121; tufted (*C. apella*), 135
Catarrhini (catarrhines), 19–32,
  55
Cebidae, 14–16
Cercopithecidae, 19, 21
Cercopithecinae, 19–28
Cheirogaleidae, 7
chimpanzee (*Pan troglodytes*), **2,**
  30, 35, **35, 40,** 42, 43, 52, **53,**
  55, 56–57, **57,** 61, 76, 84, **84,
  89,** 91, 92, 93, **98, 99,** 100–
  101, 104, **108,** 112, 113, 114,
  114, 119, 119–123, **123,** 124,
  126, **126,** 129–130, 133–134,
  135–136, **137,** 140–142, **141,**
  143–148, **148,** 154–156
Chiromyiformes, 7
Colobinae, 19, 28–33
colobus monkeys, 28–30, **29,** 46,
  61, 101, 118, 124; black
  colobus (*Colobus satanas*), 28,
  29; mantled guereza (*C.
  guereza*), 39, 61, 84; olive
  colubus (*Procolobus verus*), 28,
  30; red colobus (*Piliocolobus*
  spp.), 28, 29

Daubentoniidae, 7
domestic dog (*Canis familiaris*),
  2–3
doucs (*Pygathrix* spp.), 31
douracoulis. *See* night monkeys
drill (*Mandrillus leucophaeus*), 25,
  26–7, 64

Elaeis spp., 135
elephants, 116

false potto (*Pseudopot
  to martini*), 64

Galagonidae, 9
galagos. *See* bushbaby
gelada (*Theropithecus gelada*),
  26–28, 60
gibbons (*Hylobates* spp.), 33, 48,
  49, 50, 55, 65, 66, 75, 76, 82,
  124; northern white-cheeked
  (*H. leucogenys*), 33; white-
  handed (*H. lar*), **34, 64**
gorillas (*Gorilla* spp.), **35,** 35,
  36, 41, 43, **43,** 47, 52, 55, 56,
  58, 59, 60, 67–69, **68, 78,**
  81, 90, 92, **92,** 94, 95, **95,** 96,
  110, 112, 120–121, 154–156;
  mountain (*G. beringei
  beringei*), 93, 94, 147
great apes, 35–36, 65, 66, 96–97,
  112, 123, 131–132, 147–149
guenons (*Cercopithecus* spp.), 21,
  23–24; De Brazza's monkey
  (*C. neglectus*), 23; diana
  monkey (*C. diana*), 24

Haplorrhini (haplorrines), 5,
  10–36, 116
Hominidae, 35–36, 43, 52,
  53, 55
howler monkeys (*Alouatta* spp.),
  13, 62, 82, 124, **125**
humans (*Homo sapiens*), 3, 33,
  35, 42, 43, 50, 52, **53,** 55,
  56–57, 58, 65, 66, 70, 75, 78,
  90, 100–101, 114, 127, 139,
  145, 147–149
Hylobatidae, 33
hylobatids, 33

indri (*Indri indri*), 7, 64
Indridae, 7

langurs, 30–31; Mentawai
  (*Presbytis potenziani*), 30; pig-
  tailed (*Simias concolor*), 31,
  64; sacred (*Semnopithecus*
  spp.), 30, 151
leaf monkeys (*Trachypithecus*
  spp.), **70, 86**
Lemuridae, 7
Lemuriformes, 7
lemurs, 75; black-and-white
  ruffed (*Varecia variegata*), 8;
  mouse-lemurs (*Microcebus*
  spp.), 82; pygmy mouse-lemur
  (*Microcebus myoxinus*), 57,
  58; red-bellied (*Eulemur
  rubriventer*), **8,** 45; ring-tailed
  (*Lemur catta*), 6, **6, 80,** 81;
  ruffed (*Varecia* spp.), 90
lesser apes. *See* gibbons
Loridae, 8
Loriformes, 8
lorises (*Loris* spp.), 8–9, 46, 64

macaques (*Macaca* spp.), 24,
  46, 64, 66, 91, 100, 104, 118,
  125; Barbary (*M. sylvanus*),
  24, 43, 85, 102; Japanese (*M.
  fuscata*), 124, 137–139, **139,**
  140, 151; lion-tailed (*M.
  silenus*), **25,** 61; rhesus (*M.
  mulatta*), 78, 102, 104, 133,
  151; stump-tailed (*M.
  arctoides*), 104
mandrill (*Mandrillus sphinx*), 25,
  26–27, 64
mangabeys (*Lophocebus* spp. and
  Cercocebus spp.), 24–26
marmoset: Goeldi's marmoset
  (*Callimico goeldii*), 15. *See also*
  pygmy marmoset
maroon leaf monkey (*Presbytis
  rubicunda*), 30
Megaladapidae, 7
monkeys, 13–32
mouse-lemur (*Microcebus
  myoxinus*), 7
muriquis (*Brachyteles* spp.), 18

191

Neesia spp., 121–122, 140
New World monkeys. *See*
  Platyrrhini
night monkeys (*Aotus* spp.), 10,
  16–17, **17**
Nyctipithecidae, 16–17, 64

odd-nosed primates, 31–33
Old World monkeys. *See*
  Catarrhini
orangutans (*Pongo* spp.), 35, **39**,
  43, **51**, 52, 55, 56, 82, **83**, 84,
  90, 91, 94, 96, 104, 119–121,
  124, 130, **131**, 140, **150, 152**,
  154–156, **162**; Bornean (*P.
  pygmaeus*), 3, **4, 152**;
  Sumatran (*P. abelii*), 152
owl monkeys. *See* night monkeys

Patas monkey (*Erythrocebus
  patas*), 21, **22**, 66, **67**
Phayre's leaf monkey
  (*Trachypithecus phayrei*), 30
Pitheciidae, 17–18
Platyrrhini (platyrrhines), 13–19
potto (*Perodicticus potto*), 9, 64

Presbytis. See langurs
proboscis monkey (*Nasalis
  larvatus*), 32, **32, 48**
pygmy marmoset (*Callithrix
  pygmaea*), 14

rhesus monkeys. *See* macaques
Rodentia, 36

sakis (*Chiropotes* spp. and *Pithecia*
  spp.), 17–18
Semnopithecus. See langurs
sifakas (*Propithecus* spp.), 46, **46**
silvery leaf monkey
  (*Trachypithecus cristatus*), **31**
Simiiformes, 13–36
slow lorises (*Nycticebus* spp.), 9
snub-nosed monkeys
  (*Rhinopithecus* spp.), 31
spider monkeys (*Ateles* spp.),
  18–19, **18, 20**, 43, 49, 62, **63**
squirrel monkeys (*Saimiri* spp.),
  15, **15**, 16, 133
Strepsirrhini (strepsirrhines),
  5–10, 116

swamp monkey (*Allenopithecus
  nigroviridis*), 21

talapoins (*Miopithecus* spp.), 21
tamarin: cottontop (*Saguinus
  oedipus*), 14; golden lion
  (*Leontopithecus rosalia*), 14, 42,
  76, 81, 84, 157; lion
  (*Leontopithecus* spp.), 75, 100
tarsiers (*Tarsius* spp.), 59, 66, 100
Tarsiidae, 11–13
Tarsiiformes, 11–13
titi monkeys (*Callicebus* spp.), 18
Trachypithecus. See langurs

uakaris (*Cacajao* spp.), 18, 64

vervet monkeys (*Chlorocebus*
  spp.), 21–22, **23**, 91, **127**,
  128, 135

woolly monkeys (*Lagothrix* spp.
  and *Oreonax flavicauda*), 62
woolly spider monkeys
  (*Brachyteles* spp.), 62

SUBJECT INDEX

Page numbers in boldface indicate illustrations.

adolescence, 86–88
adoption, 93–95
Ai, 129, 133
allogrooming. *See* grooming
American Sign Language (ASL), 129
American Society of Primatologists, 161
American Zoo and Aquarium Association (AZA), 160
anatomy, 36; eyes, 40–41; hands and feet, 38–40; reproductive, 12, 37; vocal, 131–132
anthropocentrism, 71
aunting behavior, 16, 30, 85
autogrooming. *See* grooming
Azy, 129

Balikpapan Orangutan Society (BOS)—USA, 160
Balikpapan Orangutan Survival Foundation—Indonesia, 160
Baraka, 95
Belle, 144
bicornuate uterus, 7, 10
bipedalism. *See* locomotion
body size, 57–58, 68, 79–81
Bongo, 94
brachiation. *See* locomotion
brains, 41–42, 45, 116–118; absolute size, 116–117; and longevity, 117; relative size, 116–117
bushmeat, 154–156

captivity, 162–163
catarrhines, 14, 19–21
Center for Great Apes, 161
cheek pouches, 14, 19, 25, 28, 33
Chimpanzee and Human Communications Institute, 129
coccyx, 65
Columbus Zoo, 94, 159, 160
communication, 124–127; broadcast calls, 124; drumming by chimpanzees, 124; by gibbons, 33; by language, 36, 127–131; by mangabeys, 25; nonvocal, 24, 27, 28; vocalizations by vervets, 22, 128
conservation, 157–160

Conservation International, 161
counting, 132–134
culture, 139–142

Darwin, Charles, 54
deception, 142–145
Dian Fossey Gorilla Fund International, 160
diet, 59–61, 79–80
DNA (deoxyribonucleic acid), 52–54
du Chaillu, Paul Belloni, 67

emigration, 86
emotions, 146–149
endangered primate species, 151–152
epitheliochorial placentation, 7
erotic behavior, 90
evolution, 54–57

family, 71–72
fitness, 81
Flint, 147
Flo, 147
foraging, 59–61; anatomical specializations of the aye-aye for, 7; destructive, 16; for tree sap, 15
Fossey, 94
Freud, 146
friendship, 102–103, 104

Gardner, Allen, 129
Gardner, Beatrix, 129
Garner, Richard L., 112
Golden Lion Tamarin Conservation Program, 157
Goodall, Jane, **53**, 94, 118, 164
Great Ape Conservation Act, 158
grooming, 97–100
group structure, 72–75
Groves, Colin, 4

habitat destruction, 153–156
hand clasp, 142
Haplorrhini (haplorrhines), 5, 10–36, 116; reproduction, 10–11
Hayes, Cathy, 129
Hayes, Keith, 129
hemochorial placenta, 11, 12

historical perceptions, 96–97
Holoko, 95
hunting, 100–102

imitation. *See* social learning
Imo, 137–139
Indah, 94, 129, **131**
infant care, 83–86
infant development. *See* maternal care
infanticide, 79, 85
intelligence, 109; intelligence quotient (IQ), 114–116; study of, 109–110, 111–114
International Primate Protection League, 161
International Society of Primatologists, 161
International Union for Conservation of Nature and Natural Resources (IUCN), 151
Iris, 94
ischial callosities (sitting pads), 19, 21, 25, 27, 28, 33, 88

Jane Goodall Institute, 160
Japanese school of primatology, 97

Kearton, C., 146
Kejana, 95
kin recognition, 77–79
Köhler, Wolfgang, 113–114
Koshima macaques, 137–139, 140

Lamarck, Jean Baptiste, 56
language. *See* communication
language research, 128–131
Language Research Center, Georgia State University, 130
Linnaean hierarchy, 3
Linnaeus, Carolus, 1
locomotion, 45–51; bipedalism, 17, 19, 35, 44, 50–52; and body size, 80–81; brachiation, 33, 48–49, 66; of catarrhines, 21; quadrupedal, 46–47; semibrachiation, 18, 49; speed, 66; in tarsiers, 11–12; vertical clingers and leapers, 45–46
longevity, 65–66

Mandara, 95, **95**
mate, choice of, 32, 92–93
maternal care, 81–83
mating, 90–92; orgasm, 90
meat eating. *See* hunting
micromanipulation, 59
mirror, recognition of reflection
    in, 36, 44
monogamy, 75–77, 86; and body
    size, 81; in gibbons, 33; in
    golden lion tamarins, 14; in
    Mentawai langur, 30; in titi
    monkeys, 18
movement. *See* locomotion

National Zoo, 95
Nature Conservancy, 161
neoteny, 84–85; coloration in
    langurs, 30–31
New World monkeys. *See*
    platyrrhines
nocturnal lifestyle, 16–17
nursing. *See* maternal care
nut cracking, 122–123

Ohio State University, 133
Old World monkeys. *See*
    catarrhines
opposable digits, 1, 14, 24, 35,
    39–40

Partners in Conservation (PIC),
    159, 160
paternal care, 75–76, 79, 84, 85
paternity, 78–79
pestle pounding. *See* tool use
pets, primates as, 105–107
platyrrhines, 14
polyandry, 71, 76
polygamy, 71, 81, 85
prehensile. *See* tails
Primarily Primates, Inc., 161
Primate Rescue Center, 161

Primate Research Institute,
    Kyoto University, 129, 133
Primate Station, Tenerife
    Island, 112
Primate Taxonomy (Groves), 4–5
promiscuity, 79
prosimians, 5–10; definition of,
    5–6
Prussian Academy of Sciences,
    112

rank, acquisition of, 77
reconciliation, 103–105
Red List, 151
retia mirabilia, 8
Rock, 142

sanctuaries, 158–160
semibrachiation. *See* locomotion
sexual dichromatism, 33–34
sexual dimorphism: in ape teeth,
    33; in baboons, 26; in geladas,
    28; in hair pattern of great
    apes, 35; and social system, 81
sexual skin, 88–89; in baboons,
    27; lack of swelling in colobus,
    29; lack of swelling in
    guenons, 23; lack of swelling
    in platyrrhines, 14; in
    mangabeys, 25–26; minor
    swelling in gibbons, 33;
    presence in *Piliocolobus* and
    Procolobus, 29–30; presence in
    tarsiers, 12; in swamp
    monkeys, 21
Sheba, 133
sitting pads. *See* ischial callosities
skeletons, **53**
social learning, 136–139,
    139–142
social systems, 72–75; and body
    size, 81; costs and benefits of,
    72–74

speed. *See* locomotion
strength, 67–69
Strepsirrhini (strepsirrhines),
    5–10, 116; definition of, 5;
    reproduction, 7
Support for African/Asian Great
    Apes (SAGA)—Japan, 160
swelling. *See* sexual skin
swimming, 21, 32
sympathy, 145–146

tails, 43, 61–64; prehensile, 16,
    18–19, 21, 61–64; tailbone in
    humans, 65
tapetum lucidum cellulosum, 6,
    12, 16
taxonomic classification of
    primates, 1–35; explanation
    of, 1–4; among great apes,
    52–54
termite fishing, 122
Teuber, Eugen, 112–113
thumbs, 39–40; absence of in
    Atelidae, 18; absence of in
    Colobinae, 28; in great apes, 35
tool manufacture, 36, 120–123
tool use, 36, 118–120, 136–137,
    140–141; definition of, 118;
    pestle pounding, 122
Toto, 146
tumescence. *See* sexual skin

urine washing, 10, 12

Vicki, 129

Washoe, 129
World Wildlife Fund, 161

Yerkes, Robert M., 114